ANIMAL SCIENCE, ISSUES AND PROFESSIONS

T0290959

HUMAN AND ANIMAL MATING

STRATEGIES, GENDER DIFFERENCES AND ENVIRONMENTAL INFLUENCES

ANIMAL SCIENCE, ISSUES AND PROFESSIONS

Additional books in this series can be found on Nova's website under the Series tab.

Additional e-books in this series can be found on Nova's website under the e-book tab.

HUMAN REPRODUCTIVE SYSTEM - ANATOMY, ROLES AND DISORDERS

Additional books in this series can be found on Nova's website under the Series tab.

Additional e-books in this series can be found on Nova's website under the e-book tab.

ANIMAL SCIENCE, ISSUES AND PROFESSIONS

HUMAN AND ANIMAL MATING

STRATEGIES, GENDER DIFFERENCES AND ENVIRONMENTAL INFLUENCES

MISAKI NAKAMURA

AND

TAKAKO ITO

EDITORS

New York

Copyright © 2013 by Nova Science Publishers, Inc.

All rights reserved. No part of this book may be reproduced, stored in a retrieval system or transmitted in any form or by any means: electronic, electrostatic, magnetic, tape, mechanical photocopying, recording or otherwise without the written permission of the Publisher.

For permission to use material from this book please contact us:
Telephone 631-231-7269; Fax 631-231-8175
Web Site: http://www.novapublishers.com

NOTICE TO THE READER

The Publisher has taken reasonable care in the preparation of this book, but makes no expressed or implied warranty of any kind and assumes no responsibility for any errors or omissions. No liability is assumed for incidental or consequential damages in connection with or arising out of information contained in this book. The Publisher shall not be liable for any special, consequential, or exemplary damages resulting, in whole or in part, from the readers' use of, or reliance upon, this material. Any parts of this book based on government reports are so indicated and copyright is claimed for those parts to the extent applicable to compilations of such works.

Independent verification should be sought for any data, advice or recommendations contained in this book. In addition, no responsibility is assumed by the publisher for any injury and/or damage to persons or property arising from any methods, products, instructions, ideas or otherwise contained in this publication.

This publication is designed to provide accurate and authoritative information with regard to the subject matter covered herein. It is sold with the clear understanding that the Publisher is not engaged in rendering legal or any other professional services. If legal or any other expert assistance is required, the services of a competent person should be sought. FROM A DECLARATION OF PARTICIPANTS JOINTLY ADOPTED BY A COMMITTEE OF THE AMERICAN BAR ASSOCIATION AND A COMMITTEE OF PUBLISHERS.

Additional color graphics may be available in the e-book version of this book.

Library of Congress Cataloging-in-Publication Data

ISBN: 978-1-62417-085-0

Published by Nova Science Publishers, Inc. † New York

CONTENTS

PREFACE

This book presents topical research on human and animal mating strategies, gender differences and environmental influences. Topics include the mating effects on female reproductive organs and the paradigm of estrogen signaling pathways in the oviduct; sexual maturation, mating strategies and neuroendocrinology in social insects; plasticity in mating patterns of a benthic nest-holding fish related to the effects of nest-site abundance and social interactions; and human sexual strategies of short-term mating and parental control over mate choice.

Chapter 1 – Mating induces several physiological changes in the female reproductive tract independently of oocyte fertilization, which are potentially required for a successful pregnancy. These effects include modifications in the cellular and molecular mechanisms by which some steroidal hormones exert their actions in the oviduct and uterus, changes in the expression of some key molecules inside the uterus and modulation of the steroidogenic functions in the ovary. Mating impinges on the female reproductive tract with sensory stimulation, seminal fluid and sperm cells, so these effects may involve a direct interaction between the cells of the female reproductive tract and the different components of the seminal plasma and/or spermatozoa to modulate the immune system or neuroendocrine changes elucidated by the mechanical stimulation of the cervix, which indirectly affect the functioning of the cells that compose the female reproductive tract. In this chapter the authors will review the available literature on the effects of mating in the physiology of oviduct, ovary and uterus highlighting a hitherto unsuspected early, strong and broad influence of mating on the physiology of female reproductive organs. The authors will be specially centered into review of the well-defined effect of mating on the rat oviduct, where it changes the mechanism by which estradiol accelerates oviductal egg transport, from a nongenomic to a genomic pathway. This change has been named intracellular pathway shifting (IPS) and reflects a novel example of functional plasticity in well-differentiated cells induced by mating. The authors will describe that IPS involves inhibition of the conversion of estradiol to 2-methoxyestradiol, which probably protects the embryos from the deleterious effect that methoxyestradiols exert during the embryo development. IPS seems to involve changes in intraoviductal signaling mediated by cytokines TNF-α and TGFβ and is independently induced either by cervico-vaginal stimulation with a rod glass or by intrauterine insemination. Such redundancy of triggering factors suggests that IPS is an important element in the reproductive strategy.

Chapter 2 – The process of mating in social insects is unique among insect species because reproductive individuals interact with other colony members including reproductives and non-reproductives. In some species, reproductives go through intra-caste competition to monopolize the reproduction in a colony before mating. They show diverse strategies for this competition. Reproductives may undergo some physiological and behavioral changes prior to the onset of mating flight as well as dynamic changes after mating. Although the mechanisms underlying these changes differ depending on the species and sex, particular neuroendocrines appear to regulate the changes in some species. In honeybees, dopamine is likely to be involved in behavioral activation before mating. Physiological factors affecting the amount of brain dopamine may differ between the sexes. Sexual difference in dopamine regulation might result from their different social roles in the colony. The present chapter also addresses multiple mating as a queen's mating strategy. Some queens are monoandrous and mate only once, but others are polyandrous and may continue mating attempts until they mate with a sufficient number of males. Honeybee queens are likely to monitor the total amount of semen received in the previous mating to determine the timing of cessation of mating flight. By multiple mating, they increase the genetic heterogeneity in workers and enhance colony performance. Multiple mating also achieves lower intra-colony relatedness and subsequently reduces the queen-worker conflicts over sexual production.

Chapter 3 – For nest-holding animals such as fishes, abundance and spatial-temporal distribution of resources essential for reproduction (e.g. nests, breeding territories) is incredibly important for understanding the demographic and evolutionary consequences of sexual selection. In this chapter, firstly the author will review the plasticity of mating patterns of the Japanese fluvial sculpin (*Cottus pollux*) related to nest site abundance in their natural condition.

The number of eggs males obtained in their nests was significantly different between the two study sites with different nest site abundance, which may support the plasticity of their mating patterns related to nest abundance. Larger males occupied nests earlier and nesting males were larger than non-nesting males (i.e. size-dependent reproduction) in the area with a shortage of nest sites, whereas the same trend was not apparent in the area with sufficient nest site abundance, suggesting competitive exclusion among males for nest site under low nest abundance site. Also, males inhabit under shortage nest abundance site tend to have early maturity and shorter life span than males inhabit under sufficient nest abundance site. Secondly, the author will review nest choice experiments by males and subsequent mating experiments to quantify the effects of male-male competition on nest site choice and mating success of the male sculpins under both sufficient and shortage nest-abundance conditions. Contrary to the traditional prediction regarding mating pattern plasticity in animals in relation to the changes in nest abundance (i.e. mating pattern can shift from polygynous to monogamous as nest-site abundance increased), the results indicated exclusive polygynous mating patterns for the sculpin regardless of nest abundance. In this species, size-mediated dominance and aggressive behaviour of males may disrupt nest acquisition by other conspecific males, and may consequently result in extreme variation in mating success among males even under sufficient nest-abundance conditions. Finally, the author will propose several management implications for conservation of the sculpin derived from the findings of these two topics.

Chapter 4 – In the human species, individuals engage in short-term mating strategies that enable them to acquire fitness benefits from casual mates. However, because parents and children are not genetically identical, these benefits are less valuable and more costly to their parents. For this reason the latter are likely to disapprove their children engaging in casual relationships. This chapter aims to review the evidence from several studies which indicates that parents and children disagree over short-term mating strategies, with the former considering them less acceptable for the latter than the latter consider them acceptable for themselves. Moreover, parents find it more unacceptable for their daughters to engage in short-term mating than for their sons, while male children consider casual mating more acceptable than female children. Also, mothers are more disapproving than fathers of their children's short-term mating strategies, while both parents and children consider short-term mating less acceptable within marriage. The implications of these findings for interfamily conflict are also explored.

In: Human and Animal Mating
Editors: M. Nakamura and T. Ito

ISBN: 978-1-62417-085-0
© 2013 Nova Science Publishers, Inc.

Chapter 1

MATING EFFECTS ON FEMALE REPRODUCTIVE ORGANS: THE PARADIGM OF ESTROGEN SIGNALING PATHWAYS IN THE OVIDUCT

Pedro A. Orihuela[1], Alexis Parada-Bustamante[2],
Hugo Cárdenas[1], Alejandro Tapia-Pizarro[2],
Lorena Oróstica[1] and Patricia Reuquén[1,3]

[1]Laboratorio de Inmunología de la Reproducción,
Facultad de Química y Biología, Universidad de Santiago de Chile, Chile
[2]Instituto de Investigaciones Materno-Infantil, Facultad de Medicina,
Universidad de Chile, Chile
[3]Escuela de Bioquímica, Universidad Andres Bello, Chile

ABSTRACT

Mating induces several physiological changes in the female reproductive tract independently of oocyte fertilization, which are potentially required for a successful pregnancy. These effects include modifications in the cellular and molecular mechanisms by which some steroidal hormones exert their actions in the oviduct and uterus, changes in the expression of some key molecules inside the uterus and modulation of the steroidogenic functions in the ovary.

Mating impinges on the female reproductive tract with sensory stimulation, seminal fluid and sperm cells, so these effects may involve a direct interaction between the cells of the female reproductive tract and the different components of the seminal plasma and/or spermatozoa to modulate the immune system or neuroendocrine changes elucidated by the mechanical stimulation of the cervix, which indirectly affect the functioning of the cells that compose the female reproductive tract.

In this chapter we will review the available literature on the effects of mating in the physiology of oviduct, ovary and uterus highlighting a hitherto unsuspected early, strong and broad influence of mating on the physiology of female reproductive organs. We will be specially centered into review of the well-defined effect of mating on the rat oviduct, where it changes the mechanism by which estradiol accelerates oviductal egg transport, from a nongenomic to a genomic pathway. This change has been named intracellular pathway shifting (IPS) and reflects a novel example of functional plasticity in well-differentiated cells induced by mating. We will describe that IPS involves inhibition of the conversion of estradiol to 2-methoxyestradiol, which probably protects the embryos from the deleterious effect that methoxyestradiols exert during the embryo development. IPS seems to involve changes in intraoviductal signaling mediated by cytokines TNF-α and TGFβ and is independently induced either by cervico-vaginal stimulation with a rod glass or by intrauterine insemination. Such redundancy of triggering factors suggests that IPS is an important element in the reproductive strategy.

Keywords: Mating, female reproductive organs, estradiol, reproductive strategy, and cytokines

INTRODUCTION

Mating components include sensory stimulation, seminal plasma and spermatozoa. Either individually or collectively, these constituents impact the female physiology through their interaction with cells composing the female reproductive tract. It is now widely accepted that, independently of its fertilizing role, mating has physiological relevance since it affects at molecular and cellular level the functioning of reproductive organs near or beyond to the site of insemination.

The relevance of mating-associated factors on the molecular and cellular changes occurring in the female reproductive organs is clearly illustrated by the fact that both cervico-vaginal stimulation and the presence of spermatozoa can itself changes the mechanism by which estradiol (E_2) regulates oviductal egg transport, from a nongenomic mode to a genomic mode. We have denominated this shunt in E_2 pathways *intracellular path shifting* (IPS). The biological basis of IPS involves silencing of the intraoviductal E_2 non-genomic pathway at least at two levels: 1) Inhibiting the expression and activity of Catechol-O-Methyl-Transferase (COMT) therefore decreasing production of 2-Methoxyestradiol (2ME) in the oviductal cells and 2) silencing the signaling cascades downstream of 2ME in the oviductal cells.

In this chapter we review the effects of mating and associated components on ovarian and uterine physiology of some mammals and discuss the molecular and cellular mechanisms that induce IPS specifically in the rat oviduct.

MATING EFFECTS ON OVARIAN FUNCTION

In mammals is crucial that after mating, ovaries continue the production of progesterone to ensure a correct implantation and development of the embryo in the uterine cavity. For several years, it has been postulated that some mating components would play a role in this phenomenon. According to the literature, seminal plasma and cervico-vaginal stimulation provided by coitus alter ovarian physiology in order to rescue and/or improve the functioning of corpus luteum, being the importance of these components species-dependent. The seminal plasma changes the number of macrophages within the ovary, affecting the steroidogenic function of the corpus luteum; while cervico-vaginal stimulation changes the pattern of secretion of prolactin, which controls the life span of the corpora lutea.

Seminal Plasma Modifies Macrophages Population in the Ovary

The most accepted role for seminal plasma (SP) in the reproductive process in mammals is to facilitate the transport and survival of spermatozoa during their passage throughout the female reproductive tract; however, in the last years it has been shown that some molecules of SP generate dramatic changes in the female reproductive tract, which would be important for fertilization and implantation. For example, as will be shown below, some components of SP induce a like-inflammatory reaction in endometrium of sows (O'Leary et al; 2004), mares (Portus et al; 2005, Palm et al; 2008) and mice (Tremellen et al; 1998). This phenomenon would be generated by a direct interaction of some SP components with the endometrium (Robertson; 2005).

Unlike the uterus, the mammalian ovary do not have direct contact with SP after mating; however, SP is capable of inducing the recruitment of immune cells in this organ, a phenomenon that affects the processes of follicular development, ovulation and luteinization of corpus luteum in a manner dependent on the species (reviewed in Pate et al; 2010). For example, female mice mated with males without seminal vesicles showed a significantly fewer number of macrophages within corpora lutea on the day after mating compared to females mated to vasectomized or intact males (Gangnuss et al; 2004). However, no differences were observed in the ovulated ova or on the ability of corpora lutea to produce progesterone between these experimental groups, indicating that in mice the recruitment of macrophages in the ovarian tissue is not important in the changes of the steroidogenic function required after mating. As in mice, pigs that received seminal plasma transcervically had a increased number of macrophages in the ovaries (fourfold) compared to animals treated with phosphate buffered saline (O´Leary et al; 2006). Although there were not differences in the number of oocytes ovulated between these two groups, the weight of corpora lutea and serum progesterone levels were increased in pigs treated with SP (O´Leary et al; 2006), indicating than in pigs, in difference to the mice, the recruitment of macrophages in the ovary is associated with intrinsic changes in the steroidogenic ovarian function. The mechanisms involved in the recruitment of immune cells in the ovary after mating have not been identified; however, it has been postulated the existence of a semen–uterine–ovarian axis, where the recruitment of macrophages into the uterus, induced by the interaction between some components of SP with the endometrium, generates semen-induced cytokines, which are subjected to a counter-current transfer to the ipsilateral ovary, where they attract immune cells (Robertson; 2007). This idea is supported by the observation that in gilts there is a counter-current transfer from the uterus into ovarian arterial (Waberski et al; 2006).

Sensory Stimulation Reprograms Prolactin Secretion, Affecting Ovarian Function

Prolactin (PRL) is a peptide hormone produced by the anterior pituitary gland in specialized cells named lactotrophs. This hormone has several biological functions including to promote the production of milk by breast gland during lactation and to rescue the corpus luteum after mating (reviewed in Freeman et al. 2000).

The mechanisms by which PRL rescues corpus luteum have been broadly characterized in murine, where this phenomenon is crucial for a successful pregnancy. For example, female mice that do not express PRL receptor are infertile (Ormandy et al. 1997), probably because their corpora lutea suffer early regression, which could be directly related to the decrease in progesterone production that occurs in these animals following of mating (reviewed in Bole-Feysot et al. 1998).

In unmated rats, PRL has a single preovulatory surge during the estrous cycle (Butcher et al. 1974), whereas in mated rats twice-daily PRL surges are triggered within 48 hours after mating and it persists for 10–12 days (Butcher et al. 1972). This change in the pattern of PRL secretion induced by mating is required for the rescue of the corpora lutea and the factor that induces this change is the cervico-vaginal stimulation, since animals stimulated with a glass rod do not experiment corpora lutea regression and they have a prolactin serum pattern similar to that detected in mated animals (Smith et al. 1975). Although, the change in PRL secretion after mating has a crucial importance in ovary functioning, this phenomenon is ovary-independent, given that PRL surges observed after mating can be initiated by cervico-vaginal stimulation in ovariectomized rats (Freeman and Nill, 1972, Helena et al., 2009). The mechanism by which mating regulates PRL release is yet unknown but it could involve a change in the MAPK activity because we have recently shown that levels of phophoprylated MAPK are increased in the rat pituitary gland between 6-12 hours following a cervico-vaginal stimulation with a rod glass in the night of proestrus (Reuquén et al., 2012).

In rats, PRL has a dual effect on corpus luteum depending whether the animal is mated or not. In unmated rats, the PRL surge induces regression of the corpora lutea of previous cycles (luteolytic function); whereas in mated rats, the twice-daily PRL surge facilitate the development of newly formed corpora lutea (luteotrophic function). The mechanisms by which PRL exert this dual function would involve, in both cases, an increase in the number of macrophages within ovary induced by PRL, since an inhibitor of prolactin production (bromocryptine) decreased the number of macrophages in the corpora lutea in both conditions, while a PRL treatment to unmated rats increased the number of macrophages in the corpora lutea (Gaytán et al. 1997).

The cellular mechanisms by which PRL exerts their actions on corpora lutea in mated and unmated rats are not completely clear. The luteotrophic function of PRL after mating would involve an improved progesterone secretion, through suppressing 20 alpha-hydroxysteroid dehydrogenase activity in the corpus luteum (Matsuyama et al. 1990), being this action mediated by factors secreted by macrophages recruited within corpus luteum (Kirsch et al. 1981). Yamanouchi et al (1992) performed an elegant experiment that supports the relationship between macrophages and PRL to regulate progesterone secretion by luteal cells. These investigators shown that PRL induced a significantly higher level of secreted progestin in mature rat granulose cells co-cultured with splenic macrophages in comparison to those levels induced by PRL in a monoculture of mature granulose cells (Yamanouchi et al. 1992). In rats, the factor produced by macrophages involved in this phenomenon would be transforming growth factor-β (Matsuyama et al. 1990).

Regarding to the effect of PRL on luteal regression in unmated rats, this phenomenon occurs during the transition from proestrus to estrus. When the preovulatory PRL surge is blocked in rats, the recruitment of macrophages in corpora lutea and the apoptosis of luteal cells is inhibited (Gaytán et al. 1998), showing the physiological importance of PRL on these effects in murine. The mechanisms by which PRL induces luteal regression in unmated rats would involve mainly apoptosis of luteal cells (Guo et al. 1998; Gaytán et al. 2001) and it is a process dependent of the luteal cells differentiation degree (Gaytán et al. 2001).

Finally, it was shown that the profile of prolactin secretion in women also changes after intercourse (Kruger et al; 2012). If this change, induced by mating, favors a successful pregnancy in humans as in rodents, remains to be elucidated.

A Component of Seminal Plasma Induces Ovulation in New World Camelids

In some mammals, mating could induce ovulation (as for example in cows, cats, ferrets and camelids). In most of these animals, the responsible factor for this phenomenon is the cervico-vaginal stimulation provided during coitus, which activates noradrenergic neurons of the midbrain and brainstem, which then project and activate neurons located in medial basal hypothalamus which produce and secrete GnRH (reviewed in Bakker and Baum, 2000).

Interestingly, in the last years was determined that ovulation in llamas and alpacas (two South American camelids with mating-induced ovulation) is induced by seminal plasma rather than cervico-vaginal stimulation (Adams et al; 2005). In these animals, a component of seminal plasma induces ovulation by a systemic rather than a local pathway, since animals injected intramuscularly with seminal plasma had an increased ovulation rate compared to animals receiving seminal plasma by intrauterine administration (Ratto et al. 2005).

The responsible factor of this effect, named ovulation-inducing factor (OIF), was recently characterized as a protein of 14 kDa (Ratto et al. 2011). Isolated OIF administered intramuscularly to llamas induced a preovulatory LH surge followed by ovulation and corpus luteum formation (Ratto et al. 2011). At difference to cervico-vaginal stimulation, OIF induces ovulation by a mechanism not related with changes in the GnRH levels. According to recent reports, OIF would act directly on gonadotroph cells in the anterior pituitary gland to induce Luteinizing hormone release (Bogle et al; 2012).

Although the OIF purified from llamas does not induce ovulation in cattle (Tanco et al; 2012), a recent report suggests that OIF is a common component in the seminal plasma of several species of animals that are not induced ovulators, such as equines and porcine (Bogle et al. 2011). The physiological function of this molecule in these animals remains a subject of interest not elucidated.

EFFECTS OF MATING ON UTERINE FUNCTION

The Uterine Inflammatory Response to Semen

The site where semen is deposited in the female genital tract varies within mammals with species such as humans where semen is pooled in the anterior vagina close to the cervix but only sperm swim into the cervical canal toward the uterus (Sobrero and Macleod, 1962), species like mice where semen is deposited in the vagina from where is rapidly transported into the uterine cavity (Bedford and Yanagimachi, 1992; Carballada and Esponda, 1997). Some species, such as pigs, deposit semen directly into the uterine cavity bypassing the vagina (Hunter, 1981). However, regardless of the site where semen is deposited in the female genital tract, it has been described that in many mammals species that seminal fluid induces immune changes that assist in the preparation of the female reproductive tissues for pregnancy through clearance of debris and pathogens, sperm selection, and induction of immune tolerance toward the semiallogeneic embryo (Robertson, 2005; Robertson et al., 2009). One of the main features seen in many mammals after insemination is an immediate reaction elicited by insemination consisting of a rapid and dramatic increase in the number of inflammatory cells at the site where semen was deposited. In addition to this leukocytic response occurring pot-coitus, occurs an activation of innate and adaptive immune events in a similar sequence as the inflammatory response (Robertson, 2005; Robertson et al., 2009). Notably, endometrial macrophage population rises by 2-3 fold after mating (De, Choudhuri et al. 1991; Brandon, 1993).

The full effect of seminal fluid on female immune parameters is best characterized in the mice, where the cascade of events begin when proteins from seminal fluid trigger in estrogen primed cervical and endometrial epithelial cells the synthesis of several proinflammatory cytokines such as interleukin (IL)-6, IL-8, macrophage chemotactic protein-1, granulocyte-macrophage colony-stimulating factor (GM-CSF) and many other chemokines (Sanford, De et al. 1992; Robertson et al., 1996). The increased release of proinflammatory cytokines acts as chemoatractant that induces an influx of macrophages, dendritic cells, and granulocytes into the subepithelial stromal tissue of mice uterus. Later on, at the beginning of the acute phase response, a great mass of neutrophils migrates trespassing the epithelial lining toward the uterine lumen (De, Choudhuri et al., 1991; McMaster et al., 1992; Robertson et al., 1996). This transient response decline while proinflammatory cytokine production decreases driven by rising levels of circulating progesterone (Robertson et al., 1996). In addition, seminal fluid antigens activate and expand inducible regulatory T cell populations through a process of cross-presentation by female dendritic cells in lymph nodes draining the genital tract (Moldenhauer et al., 2009; Robertson et al., 2009). Subsequently, this T cell population migrates into the endometrium to mediate immune tolerance of the conceptus during the implantation process (Guerin et al., 2011).

In the horse, where semen is deposited directly in the uterus, natural service or artificial insemination results in a uterine inflammation characterized by an influx of polymorphonuclear neutrophils (PMNs) into the uterine lumen (Troedsson et al., 1990). This reaction results in an inflammatory reaction with phagocytosis of spermatozoa and bacteria in the uterine lumen and the release of PGF2α, which causes myometrial contractions, required to clear the uterus via the cervix or lymphatic vessels (Troedsson et al., 1993). Within 36-48 h the inflammation is cleared from the uterus (Katila, 1995) providing an adequate environment for embryo implantation. A similar inflammatory response has been described in sow uterus after insemination. An influx of large populations of neutrophils to the uterine lumen takes place few hours after insemination (Lovell and Getty, 1968; Claus, 1990; Rozeboom et al., 1998) while in the endometrial stromal compartment granulocytes, lymphocytes, macrophages and dendritic cells are recruited (Bischof et al., 1994; Bischof et al., 1995; Engelhardt et al., 1997). Both seminal plasma and spermatozoa cooperate in triggering this immune response since the vasectomized boars ejaculate containing seminal plasma only, increases the population of endometrial macrophages (Bischof et al., 1994; O'Leary et al., 2004) whereas the insemination with a washed spermatozoa suspension induces a higher influx of neutrophils toward the uterine lumen than seminal plasma or semen (Rozeboom et al., 1998). Similarly as for other mammals, in humans there is an influx of neutrophils into the cervical canal (Pandya and Cohen 1985; Thompson et al., 1992) as well as a robust recruitment of macrophages, dendritic cells, and memory T cells (Sharkey et al., 2012). In addition, seminal plasma participates in the recruitment of CD56[bright] NK cells into human endometrium (Kimura et al., 2009) and elicits the expression of proinflammatory cytokines and chemokines, supporting the hypothesis that seminal fluid may act as a regulator of uterine immune cells and immune tolerance at the feto-maternal interphase.

The leukocytic response occurring pot-coitus has been also described in rabbits, sheeps, horses and other mammals (Lehrer et al., 1988; McMaster et al., 1992; Kaeoket et al., 2003; O'Leary et al., 2004; Scott et al., 2006) is believed to have an important role in the reproductive process since not only confer a survival advantage to the spermatozoa within the hostile female reproductive tract, but also participate in the induction of type-2 immune response predominance at the feto-maternal interface. In mice, embryo transfer to pseudopregnant recipients without exposure to male fluids results in a greater fetal loss and abnormalities than recipients exposed to seminal plasma by mating to vasectomized males (Watson et al., 1983). Likewise, implantation rates and fetal growth are significantly increased in recipients rats inseminated before embryo transfer (Carp et al., 1984). In sows, the administration of seminal plasma before natural and artificial mating enhance reproductive performance by increasing farrowing rate and litter size (Flowers and Esbenshade, 1993) apparently by increasing the conception rate (Mah, et al., 1985). However, it has been documented that beneficial effects of seminal plasma or heat-killed semen persist into following estrous cycles (Murray et al., 1983; Murray et al., 1986; Flowers and Esbenshade, 1993). In humans, the deposition of semen in the high vaginal area improves the pregnancy rate in an IVF program (Bellinge et al., 1986) as well as sexual intercourse during the peri-transfer period of an IVF cycle had the potential to increase the likelihood of successful early embryo implantation and development (Tremellen et al., 2000). In addition, a previous exposure to seminal fluid antigens is associated to alloimmunization to HLA antigens from the male partner (Peters et al., 2004) and reduced incidence of complications for pregnancy including preeclampsia and fetal growth disturbances (Kho et al., 2009).

Immunological Mediating Factors in Seminal Fluid

The molecular mediators contained in seminal fluid that signal to epithelial cells lining the female reproductive tract and modulate gene expression, leukocyte recruitment and activation of the inflammatory response seems to be both sperm and seminal plasma, depending on the specie. In mice, it has been shown that activation and expansion of female lymphocyte populations occur after mating, and is triggered by constituents of seminal plasma derived from the seminal vesicle glands (Robertson et al., 1996) and the prostate in humans (Lee et al., 1999). Amongst the components within seminal fluid, transforming growth factor-β (TGFβ) has been identified as the principal trigger for the maternal immune responses, which is synthesized in male accessory glands, particularly in the seminal vesicle in rodents (Tremellen et al., 1998) and the prostate in humans (Lee et al., 1999); and secreted into the fluids that contribute to semen. TGFβ in seminal fluid stimulates uterine epithelial cells to release various kinds of cytokines and induces the predominance of T helper-2 (Th2) cells in mice (Tremellen et al., 1998). Interestingly, most of the TGFβ present in seminal fluid is in the latent form and becomes active in the female genital tract after insemination by the action of plasmin and other enzymes (Tremellen et al., 1998). In humans, although it is clear that exposure to seminal fluid induces a maternal immune response (Pandya and Cohen, 1985), the molecular identity responsible these actions has not been clearly defined. TGFβ cytokines have been implicated as well (Robertson et al., 2002) since a TGFβ neutralizing antibody reverses the stimulatory action of seminal plasma on the expression of chemotactic cytokines in cervical and endometrial epithelial cell cultures, whereas recombinant TGFβ resembles the effect of seminal plasma (Tremellen et al., 1998). In addition, it is likely that chemokines present in the seminal plasma such as IL-8, CCL2, CCL3, CCL4, and CCL5; contribute to the leukocyte recruitment into the cervical epithelium and stromal tissue (Maegawa et al., 2002; Politch et al., 2007). Human seminal plasma also has powerful immunosuppressive properties containing high concentrations of various effectors such as the own TGFβ (Nocera and Chu 1993; Robertson et al., 2002), the soluble p55 tumor necrosis factor-α (TNF-α) receptor, prostaglandin E2 (PGE2), 19-hydroxi prostaglandin E (19-OH-PGE2), spermine and complement inhibitory factors (Denison et al., 1999; Robertson 2005).

In pigs has been suggested that the interplay between different fractions from sperm and seminal plasma and with the female genital tract cells, reduce the intensity and duration of the uterine inflammatory response (Rozeboom et al., 1999; Rozeboom et al., 2001; Rozeboom et al., 2001). Nonetheless, the specific molecular mediators of the maternal immune response have yet to be determined. High concentrations of cytokines from the TGFβ family have been detected in boar seminal fluid (O'Leary et al., 2002) and seem to have immunosuppressive activity (Claus, 1990). However its regulatory contribution the uterine inflammatory response in gilts has not been proven and recombinant TGFβ1 administered during artificial insemination cycles did not improve implantation rate at day 80 of pregnancy (Rhodes et al., 2006). In Pigs, estrogens appear to be another mediator of seminal fluid effects in the female genital tract. Boar seminal plasma contains abundant estrogens and the transcervical administration of estrogen improves farrowing rate (Flowers and Esbenshade, 1993) and resemble some features of the physiological response to insemination (Claus, 1990). In addition, seminal prostaglandins and oxytocin may play a role since their administration seems to elicit increases in farrowing rate and litter size (Flowers and Esbenshade, 1993).

Effects of Mating on Uterine Contractility

Copulation has long been known to stimulate myometrial activity, playing an important role in sperm transport (Hartman and Ball, 1930; Rossman, 1937) since sperm motility is not required for the rapid transport of sperm into the oviducts in cows, pigs, or rabbits, as dead sperm also have been found in the oviducts of these species (Overstreet and Tom, 1982). An increased frequency of uterine contractions after mating has been described in several species including cow (Vandemark and Hays, 1954), rabbit (Fuchs, 1972), rat (Toner and Adler, 1986) and human (Fox et al., 1970). Copulation involves vagino-cervical stimulation and semen deposition within female genital tract at ejaculation; and both stimuli have shown to induce effects on uterine contractility.

Effects of Tactile Stimulation

Video-laparoscopic studies in rats showed significant changes in the contractility patterns of the uterine horns after mating and mechanical stimulation of the cervix sufficient to induce pseudopregnancy increases the frequency of weak peristalsis, partially mimicking the changes in myometrial activity induced by copulation (Toner and Adle,r 1986; Crane and Martin, 1991). In the rat, mating produces both immediate and delayed effects on myometrial activity (Toner and Adler, 1986). The immediate effect appears during the period of active mating where uterine contractile activity doubles its activity and then briefly returns to pre-mating activity levels after ejaculation. Five minutes after ejaculation the delayed effect takes place, and is characterized by an increased activity when baseline contractility has been low, but decreased when baseline has been fast. This effects last from 30 min to several hours and are triggered by vaginocervical stimulation since artificial mechanical stimulation produce both immediate and delayed effects on uterine contractility. In addition, mating increases several-fold propagating circular contractions in the uterus but in both cranial and caudal directions (Crane and Martin, 1991). Caudally directed peristalsis would be expected to transport sperm away from the uterotubal junction. However it is speculated that back and forth flow directed by myometrial contractions may provide fresh waves of sperm to the uterotubal junction. Uterine distension due to uterine fluid volume also affects uterine contractility since distensible balloons inserted into rat uterus showed that increasing and decreasing intrauterine volume increased and decreased respectively myometrial activity (Toner and Adler, 1985).

In the sow, an enhancement of uterine contractility has been shown after electrical stimulation of the cervix (Pitjkanen and Prokofjev, 1964). In addition, the stimulation of the cervix with an insemination pipette depositing 100ml of saline stimulates uterine contractility (Claus et al., 1989) whereas the intrauterine infusion of a similar saline volume avoiding cervical stimulation does not enhances uterine activity (Langendijk et al., 2002). Oxytocin exerts peripheral contractile effects on the uterus after its release into the systemic circulation (Silverman and Zimmerman, 1983), however the insertion of a catheter or manipulation of a transcervical catheter does not induce oxytocin release (Claus and Schams, 1990; Langendijk et al., 2003) suggesting that cervical stimulation may modulate myometrial contractility though an adrenergic or cholinergic pathway. Both cholinergic and adrenergic receptors have been found in swine myometrium, mostly detected in circular and longitudinal muscle layers respectively (Taneike et al., 1991).

Effects of Seminal Fluid on Uterine Activity

Seminal plasma has shown to have more clear effects on uterine contractility than cervical vaginal stimulation, which have been described in several mammal species. In rabbits, inseminated non-motile spermatozoa suspended in saline are not subjected to fast transport to the oviduct. Whereas non-motile sperm suspended in seminal plasma and inseminated, are subjected to rapid transport to the oviducts (Overstreet and Tom, 1982). The boar sperm population reaching the oviducts 4 h after artificial insemination is greater when sperm are thawed in seminal plasma than buffer solution (Einarsson and Viring, 1973); suggesting that seminal plasma has a role in facilitating sperm transport and/or increasing their survival rate across the female genital tract. In fact, boar seminal plasma has been shown to stimulate uterine motility *in vitro* (Einarsson and Viring 1973). Removal of accessory reproductive glands from male rats showed that induction of myometrial activity required constituents from the vas deferens, seminal vesicles and coagulating glands (Ventura and Freund, 1973; Crane and Martin, 1991).

Candidate mediators for eliciting the contractile effects in the uterus include oxytocin (Vandemark and Hays, 1954), prostaglandins (Ventura and Freund, 1973) and oestrogens (Claus, 1990). Oxytocin has been traditionally recognized as a hormone with actions on female reproductive function such as parturition and milk ejection. The pregnant uterus is an important target organ for oxytocin, which is a potent uterotonic and is used clinically for induction of labour. However oxytocin has been identified in the mammalian prostate and in seminal fluid (Nicholson, 1996; Watson et al., 1999) and could induce contractile effect of seminal fluid in the uterus. Prostaglandins have shown a clearer mediator effect for the contracting and relaxing properties of human and ram seminal plasma (von Euler, 1936; Eliasson, 1959). In fact, Von Euler (1936) isolated them for first time from human and ram seminal plasma and named it prostaglandin, evidencing that they induced marked stimulatory effect on uterine smooth muscle. Semen contains high amounts of different prostaglandins (Harper, 1988), which have clear effects on the contractility of the uterus (Gottlieb et al. 1991; Portus et al. 2005). In the pig, estrogen content of seminal plasma seems to mediate uterine contractility (Claus, 1990). However the mechanism by which estrogens in seminal plasma exert its effects on uterine activity is mediated by the release of prostaglandins produced by the own uterus. Upon swine insemination, the estrogens in the ejaculate induce an immediate release of prostaglandin by the endometrium (Claus 1990). In fact, the infusion of both estrogens (Claus, 1990) and prostaglandin (Einarsson and Viring, 1973; Langendijk et al., 2002) into sow uterus increases uterine motility.

ROLE OF MATING ON THE E_2 SIGNALING PATHWAY IN THE RAT OVIDUCT

It is widely known that E_2 regulates the physiology of their target organs through intracellular receptors that acts as transcription factors to control the synthesis/degradation of specific mRNA and proteins (Welboren et al., 2009). Apart of this classical E_2 genomic pathway recently it has been accumulated a lot of evidence that some effects of E_2 on their target cells is by a non-genomic mechanism since they are insensitive to transcription or translation inhibitors (reviewed in Lössel and Wheling, 2003).

In unmated rats, E_2 administration accelerates oviductal egg transport through an intraoviductal non-genomic pathway, but when mating (coitus) occurs the E_2 non-genomic pathway that accelerates oviductal transport shift away to a genomic pathway (Orihuela et al. 2001). This shunt in E_2 pathways we have now denominated *intracellular path shifting* (IPS; see Figure 1) and involve silencing of an E_2 non-genomic signaling pathway in the oviduct. The E_2 non-genomic pathway that accelerates egg transport involves sequential activation of cAMP-PKA-PLC-IP3 signaling cascades in the oviductal cells of unmated rats (Orihuela et al. 2003, 2006;). It is emphasized that most of the reports describing E_2 non-genomic actions of E_2 are focused to determine changes in the activity or level of specific molecules such as MAPK, cAMP, IP3, etc (Farhat et al., 1996; Le Mellay et al. 1997; Singh and Gupta, 1997; Qui et al., 2003) while we show that an E_2 non-genomic action is crucial to regulate a complex physiological phenomenon (egg transport) accomplished by an entire organ (oviduct) that is formed by different cell types (Orihuela et al., 2006, 2009). On the other hand, the E_2 genomic pathway involves changes in the oviductal protein synthesis (Orihuela and Croxatto, 2001) associated to Connexin 43 (Ríos et al. 2007), Endothelin (Parada-Bustamante et al. 2012) and Calbindin 9 kDa (Rios et al., 2011) signaling routes. Interestingly, both E_2 pathways require participation of estrogen receptors (ER) although the non-genomic pathway operates activating ER-α and ER-β while the genomic only requires ER-α (Orihuela et al. 2009). In this context, ER-α and ER-β are localized in association with plasma membrane, cytoplasm and nucleus of the oviductal cells that is compatible with genomic and non-genomic action of estrogens in the oviduct (Orihuela et al., 2009). Altogether, these findings highlight a novel and hitherto influence of mating-associated signals on the intraoviductal E_2 signaling and reveal potential new regulatory mechanisms to the action of E_2 on its target cells.

It is known that E_2 is metabolized by sequential steps of hydroxylation and methylation principally mediated by enzymes as cytocromes p450 1A1 and 1B1 (CYP1A1 and 1B1, Paria et al., 1990) and Catechol-O-Methyl Transferase (COMT; Watanabe et al., 1991). Estradiol is hydroxilated to catecholestrogens 2 or 4-hydroxyestradiol by CYP1A1, 1B1 enzymes and then 2-hydroxyestradiol is O-methylated by COMT enzyme generating 2-methoxyestradiol (2ME). We have now found unequivocal evidence that COMT and its product 2ME are associated to IPS.

Intraoviductal administration of a selective inhibitor (OR486) of the enzyme COMT blocked the effect of E_2 on oviductal egg transport in unmated, but not in mated rats suggesting that conversion of E_2 to its metabolite 2-methoxyestradiol (2ME) is one of the underlying mechanisms of IPS (Parada-Bustamante et al. 2007). Furthermore, decreased activity and mRNA level of COMT is observed in the oviduct of mated rats compared with those unmated (Parada-Bustamante et al. 2007) On other hand, 2ME accelerates oviductal egg transport in unmated rats by a non-genomic action that requires activation of ER (Parada-Bustamante et al. 2007, 2010) while in mated rats 2ME did not affect embryo transport according with our previous finding that OR486 did not block the effect of E_2 on embryo transport (Parada-Bustamante et al. 2010). In addition, E_2 increased cAMP level in the oviduct of unmated rats, but not in mated rats and an activator of adenylyl cyclase accelerates ovum transport only in unmated rats (Parada-Bustamante et al. 2010). These findings clearly indicate that before mating an E_2 non-genomic signaling pathway that involves the cAMP-PKA-PLC-IP3 cascades and dependent of the COMT activity and generation of 2ME operates in the oviduct. After mating, this intraoviductal E_2 non-genomic pathway is shut down at least at two levels: 1) Inhibiting expression and activity of COMT therefore decreasing production of 2ME in the oviductal cells and 2) silencing the signaling cascades downstream of 2ME in the oviductal cells. These results show for the first time a physiological role of 2ME mediated by ER in the female reproductive tract (Parada-Bustamante et al. 2007, 2010; Orihuela et al., 2009) and indicate that E_2 and its metabolites could interact to activate intracellular signaling responsible of the non-genomic actions of estrogens.

The physiological relevance of IPS is not yet clearly determined but it is probable that decreased production of 2ME induced by mating protects to early embryos of the deleterious effects that methoxyestradiols produce on the preimplantation embryo development (Lattanzi et al. 2003). The fact that 2ME mediate an E_2 non-genomic action indicates a new mode by which the biological activity of E_2 is modulated and represent a paradigm on how 2ME acts on its target cells.

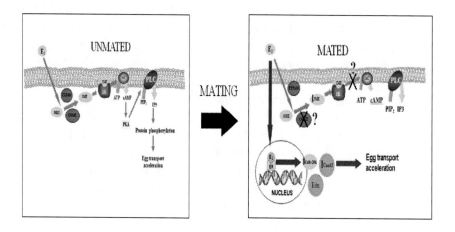

Figure 1. Tentative model that describes the intracellular path shifting of the E_2 signaling that regulates egg transport in the rat.

Role of TNF-α in IPS

As previously stated mating is able to induce an inflammatory and immune response in the female reproductive organs and thus we found that administration of an immunosupressor Ciclosporin A blocked IPS induced by sperm cells (Parada-Bustamante and Orihuela, 2002) showing the relevance of the immune system in the occurrence of IPS. Furthermore, estrogens are associated to inflammation (for detail see review of Straub, 2007). In addition, previous bioinformatics analysis of the COMT promoter has revealed binding sites for transcription factors associated to cytokines signaling including TNF-α and TGFβ. Now we are going to describe new relevant findings on the participation of TNF-α as an underlying mechanism for IPS.

We found that in the female rat, basal levels of TNF-α were higher in the oviductal fluid than in the serum both in mated and unmated rats suggesting a local activity of this cytokine in the female reproductive tract. Furthermore, TNF-α dramatically increased in the oviductal fluid at 3 hours following of mating with a fertile male whereas TGFβ level were not affected (Orostica et al., 2009). Thus, mating increase release of TNF-α in the oviduct, but not TGFβ indicating a differential effect of mating on the secretion of these cytokines in the rat oviduct.

The mammalian oviduct is a tubular organ mainly composed of an intrinsic layer smooth muscle fiber, the myosalpinx and an innermost highly folded mucosa (epithelial and stromal cells), the endosalpynx. Along of the oviduct exist differences in the amount of smooth muscle and the degree of folding and cellular composition of the mucosa (reviewed in Croxatto, 1996). Probably, IPS could entirely occur in a cell phenotype or that in a cell type E_2 non-genomic pathway will be switch off while in other cell types E_2 genomic pathway will be switch on. In order to explore in which oviductal cells IPS occurs we have examined the immunolocalization of the TNF-α receptors, TNFR1A, TNFR1B as well as the TGFβ receptor, TGFB3R. Surprisingly, mating not affect cellular distribution of TNF-α and TGFβ receptors in both oviductal layers. We postulate that IPS could be associated to changes in the release of TNF-α instead of changes in the level and distribution of their receptors in the oviduct, activating intraoviductal signaling to induce IPS.

Since both TNF-α receptors are mainly localized in the epithelial cells we also postulate that mating-induced IPS occurs in this cellular type of the oviduct. Thus, we have found that E_2 increase the cAMP levels in primary cultures of rat oviductal epithelium and this effect is blocked by Actinomicyn D and OR486 showing that the E_2 non-genomic pathway that accelerates egg transport is present in the epithelial cells. This reinforces the notion that the silencing of the E_2 non-genomic pathway induced by mating-associated signals (i.e. TNF-α) could occur in the epithelial cells of the oviduct. However, we have found that intraoviductal administration of an NF-κB activity inhibitor (Ammonium pyrrolidinedithiocarbamate, PDTC; Fröde and Calixto, 2000; Snyder et al., 2002) is not able to revert the effect of TNF-α on 2ME-induced egg transport acceleration (Oróstica et al., 2009) suggesting that TNF-α could shut down the E_2 non-genomic pathway activating intraoviductal signaling independently of the activation of its canonical transcription factor NF-κB. Probably, others transcription factors associated to COMT promoter could participate in the TNF-α signaling. Our previous bioinformatics analysis has revealed that transcription factors, as the member of the Family FKHD (FOXO1A) and GATA (GATA-3) are also associated to COMT promoter (see tentative model in Figure 2). This highlights a new research line not only to describe a new physiological function of TNF-α in the female reproductive tract but also help to elucidate a new molecular mechanism by which TNF-α acts on their target cells.

Regulation of the Activity of Estrogen Receptors in the Plasma Membrane

The E_2 non-genomic as well as genomic signaling requires activation of estrogen receptors (ER) and that ER-α and ER-β would be associated to plasma membrane, cytoplasm and nucleus of epithelial cells of the oviduct of unmated and mated rats in a localization compatible with genomic and non-genomic action of E_2 (see Orihuela et al., 2009). However, mating did not increase the number of receptors in the nucleus suggesting that IPS occurs independently of changes in the sub-cellular distribution of both ER isoforms (Orihuela et al., 2009).

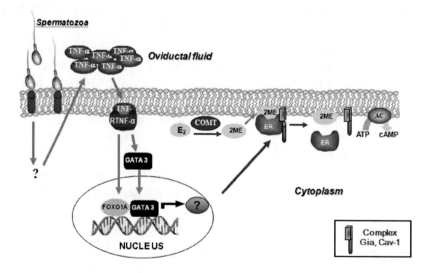

Figure 2. Tentative model that describes how the cytokine TNF-α release induced by sperm cells is able to shut down the E_2 signaling pathway in the rat oviduct.

We have found that E_2 or 2ME treatment induced co-immunoprecipitation between ER-α and Gαi protein, but not PI3K, in the plasma membrane of the oviducts of unmated rats while ER-α did not co-immunoprecipitated with Gαi or PI3K in the oviduct of mated rats after 2ME treatment. This suggest that interaction between ER-α and Gαi in the oviductal plasma membrane mediates intraoviductal E_2 non-genomic signaling and that mating-induced molecules (i.e. TNF-α) shut down this non-genomic pathway disrupting this interaction (figure 2).

Participation of Sperm Cells in IPS

Contrary to the female rabbit or cat, ovulation in the rat occurs independently of mating (Niswender et al., 2000) indicating that events related to the ovulatory process are not relevant to induce IPS. Mating impinges to the female reproductive tract principally with 3 components: sensory stimulation (Erskin, 1995), seminal fluid (Sharkey et al., 2007) and sperm cells (Suarez and Pacey, 2006). Previously, our group has found that cervico-vaginal stimulation with a rod glass (Peñarroja-Matutano, 2007) or intrauterine insemination (Parada-Bustamante et al. 2003) can independently induce IPS suggesting that this redundancy in an species in which mating invariably conduct to pregnancy must be an important element in the reproductive strategy.

In several species, sperm cells can induce changes in the gene or protein expression profile in the oviduct. In equine and bovine, two-dimensional electrophoresis gels revealed *de novo* protein synthesis in cultured oviductal cells following to exposition to sperm cells, however none of the proteins were absolutely identified (Ellington et al., 1993; Thomas et al., 1995). More recently, Fazeli et al (2004) utilized a microarray analysis to determine the effect of spermatozoids provided by mating on the transcriptome profile in the mice oviduct. They found that 6 hours after mating occurred a 2% change in oviductal gene transcription. Interestingly, a group of mating-regulated genes in the oviduct were associated with cellular responses involving cytokines and inflammation. Furthermore, our group has found that a total of 17 transcripts showed greater than 2-fold changes 6 h after mating in the rat oviduct (Parada-Bustamante et al., 2007). In addition, we have found that the oviductal gene expression pattern induced by E_2 is different between unmated and mated rats (Parada-Bustamante et al., 2010). Altogether support the view that sperm-recognition mechanisms are present in the mammalian oviduct that leads to modulation of changes in the oviductal gene expression. This reinforces our proposition that the presence of sperm cells in the female genital tract induces changes in the expression of some relevant genes that activate TNF-α signaling to underlie the silencing of the intraoviductal E_2 non-genomic pathway.

ACKNOWLEDGMENTS

Part of this work was supported by FONDECYT 1110662 to PAO, 11110457 to APB, and 11100443 to AT.

REFERENCES

Adams G. P., M. H. Ratto, W. Huanca, J. Singh. Ovulation-inducing factor in the seminal plasma of alpacas and llamas. *Biol Reprod*, 2005, 73, 452-7.

Bakker J., M. J. Baum. Neuroendocrine regulation of GnRH release in induced ovulators. *Front Neuroendocrinol*, 2000, 21, 220-62.

Bedford, J. M. and R. Yanagimachi. Initiation of sperm motility afte mating in the rat and hamster. *J Androl*, 1992, 13, 444-449.

Bellinge, B. S., C. M. Copeland, et al. The influence of patient insemination on the implantation rate in an in vitro fertilization and embryo transfer program. *Fertil Steril*, 1986, 46, 252-256.

Bischof, R. J., C. S. Lee, et al. Inflammatory response in the pig uterus induced by seminal plasma. *J Reprod Immunol*, 1994, 26, 131-146.

Bischof, R. J., M. R. Brandon, et al. Cellular immune responses in the pig uterus during pregnancy. *J Reprod Immunol*, 1995, 29, 161-178.

Bogle O. A., M. H. Ratto, G. P. Adams. Evidence for the conservation of biological activity of ovulation-inducing factor in seminal plasma. *Reproduction*, 2011, 142, 277-83.

Bogle O. A., M. H. Ratto, G. P. Adams. Ovulation-inducing factor (OIF) induces LH secretion from pituitary cells. *Anim Reprod Sci*, 2012 (In press)

Bole-Feysot C., V. Goffin, M. Edery, et al. Prolactin (PRL) and its receptor: actions, signal transduction pathways and phenotypes observed in PRL receptor knockout mice. *Endocr Rev*, 1998, 19, 225-68.

Brandon, J. M. Leucocyte distribution in the uterus during the preimplantation period of pregnancy and phagocyte recruitment to sites of blastocyst attachment in mice. *J Reprod Fertil*, 1993, 98, 567-576.

Butcher R. L., N. W. Fugo, W. E. Collins. Semicircadian rhythm in plasma levels of prolactin during early gestation in the rat. *Endocrinology*, 1972, 90, 1125-7.

Butcher R.L., W. E. Collins, N. W. Fugo. Plasma concentration of LH, FSH, prolactin, progesterone and estradiol-17beta throughout the 4-day estrous cycle of the rat. *Endocrinology*, 1974, 94, 1704-8.

Carballada, R. and P. Esponda. Fate and distribution of seminal plasma proteins in the genital tract of the female rat after natural mating. *J Reprod Fertil*, 1997, 109, 325-335.

Carp, H. J., D. M. Serr, et al. Influence of insemination on the implantation of transferred rat blastocysts. *Gynecol Obstet Invest*, 1984, 18, 194-198.

Claus, R. Physiological role of seminal components in the reproductive tract of the female pig. J Reprod Fertil Suppl, 1990, 40: 117-131.

Claus, R. and D. Schams. Influence of mating and intra-uterine oestradiol infusion on peripheral oxytocin concentrations in the sow. *J Endocrinol*, 1990, 126, 361-365.

Claus, R., F. Ellendorff, et al. Spontaneous electromyographic activity throughout the cycle in the sow and its change by intrauterine oestrogen infusion during oestrus. *J Reprod Fertil*, 1989, 87, 543-551.

Crane, L. H. and L. Martin. Postcopulatory myometrial activity in the rat as seen by video-laparoscopy. *Reprod Fertil Dev*, 1991, **3**, 685-698.

Croxatto H. B. Gamete transport. In: Reproductive Endocrinology, Surgery and Technology, Adashi EY, Rock JA, Rosenwaks Z. (eds). Lippincott-Raven Publishers, Philadelphia, 1996; 385-402.

De, M., R. Choudhuri, et al. Determination of the number and distribution of macrophages, lymphocytes, and granulocytes in the mouse uterus from mating through implantation. *J Leukoc Biol*, 1991, 50, 252-262.

Denison, F. C., V. E. Grant, et al. Seminal plasma components stimulate interleukin-8 and interleukin-10 release. *Mol Hum Reprod*, 1999, 5, 220-226.

Einarsson, S. and S. Viring. Distribution of frozen-thawed spermatozoa in the reproductive tract of gilts at different time intervals after insemination. *J Reprod Fertil*, 1973a, 32, 117-120.

Einarsson, S. and S. Viring. Effect of boar seminal plasma on the porcine uterus and the isthmus part of oviducts in vitro. *Acta Vet Scand*, 1973b, 14, 639-641.

Ellington J. E., G. G. Ignotz, B. A. Ball, V. N. Meyers-Wallen, W. B. Curie. De novo protein synthesis by bovine uterine tube (oviduct) epithelial cells changes during co-culture with bull spermatozoa. *Biol Reprod*, 1993, 48, 851-856.

Engelhardt, H., B. A. Croy, et al. Role of uterine immune cells in early pregnancy in pigs. *J Reprod Fertil Suppl*, 1997, 52, 115-131.

Erskin, M. Prolactin release after mating and genitosensory stimulation in females. *Endocr. Rev*, 1995, 16: 508-528.

Farhat M. Y., S. Abiyounes, B. Dingaan, et al. Estradiol increases cyclic adenosine monophosphate in rat pulmonary vascular smooth muscle cells by a nongenomic mechanism. *J Pharmacol Exp Ther*, 1996, 276, 652-657.

Fazeli A, N. A. Alfara, M. Hubank, W. Holt. Sperm-induced modification on the oviductal gene expression profile after natural insemination on mice. *Biol Reprod*, 2004, 71, 60-65.

Flowers, W. L., K. L. Esbenshade. Optimizing management of natural and artificial matings in swine. *J Reprod Fertil Suppl*, 1993, 48: 217-228.

Fox, C. A., H. S. Wolff, et al. Measurement of intravaginal and intra-uterine pressures during human coitus by radiotelemetry. *J. Reprod. Fertil*, 1970, 22, 243-251.

Freeman M. E., B. Kanyicska, A. Lerant, G. Nagy. Prolactin: structure, function, and regulation of secretion. *Physiol Rev*, 2000, 80, 1523-631.

Freeman M. E., J. D. Neill. The pattern of prolactin secretion during pseudopregnancy in the rat: a daily nocturnal surge. *Endocrinology*, 1972, 90, 1292–4

Fröde-Saelh T. S., J. B. Calixto. Synergistic antinflammatory effect of NF-kB inhibitors and steroidal or non steroidal antiinflammatory drugs in the pleural inflammation induced by carrageenan in mice. *Inflamm Res*, 2000, 49, 330-337.

Fuchs, A. R. Uterine activity during and after mating in the rabbit. Fertil Steril, 1972, 23, 915-923.

Gangnuss S, M. L. Sutton-McDowall, S. A. Robertson, D. T. Armstrong. Seminal plasma regulates corpora lutea macrophage populations during early pregnancy in mice. *Biol Reprod*, 2004, 71, 1135-4

Gaytán F., C. Morales, C. Bellido, et al. Role of prolactin in the regulation of macrophages and in the proliferative activity of vascular cells in newly formed and regressing rat corpora lutea. *Biol Reprod*, 1997, 57, 478-86.

Gaytán F., C. Bellido, C. Morales, et al. Both prolactin and progesterone in proestrus are necessary for the induction of apoptosis in the regressing corpus luteum of the rat. *Biol Reprod*, 1998, 59, 1200-6.

Gaytán F, C. Bellido, C. Morales, et al. Luteolytic effect of prolactin is dependent on the degree of differentiation of luteal cells in the rat. *Biol Reprod*, 2001, 65, 433-41.

Gottlieb, C., E. Andersson, et al. The effect of 19-hydroxy prostaglandins on the human myometrium in vitro. *Prostaglandins*, 1991, **41**, 607-613.

Guo K., V. Wolf, A. M. Dharmarajan, et al. Apoptosis-associated gene expression in the corpus luteum of the rat. *Biol Reprod*, 1998, 58, 739–46.

Guerin, L. R., L. M. Moldenhauer, et al. Seminal fluid regulates accumulation of FOXP3+ regulatory T Cells in the preimplantation mouse uterus through expanding the FOXP3+ cell pool and CCL19-mediated recruitment. *Biol Reprod*, 2011, 85, 397-408.

Harper, M. J. K. Gamete and zygote transport. The Physiology of Reproduction. 1988, E. Knobil and J. Neill. Nwe York, Raven Press. 1: 103-134.

Hartman, C. G. and J. Ball. On the almost instantaneous transport of spermatozoa through the cervix and the uterus in the rat. *Proc. Soc. Exp. Biol. Med*, 1930, 28, 312-314.

Helena C. V., D. T. McKee, R. Bertram, A. M. Walker, M. E. Freeman. The rhythmic secretion of mating-induced prolactin secretion is controlled by prolactin acting centrally. *Endocrinology*, 2009, 150, 3245-51

Hunter, R. H. Sperm transport and reservoirs in the pig oviduct in relation to the time of ovulation. *J Reprod Fertil*, 1981, 63, 109-117

Katila, T. Onset and duration of uterine inflammatory response of mares after insemination with fresh semen. *Biol. Reprod. Monogr*, 1995, 1, 515-517.

Kaeoket, K., E. Persson, et al. Influence of pre-ovulatory insemination and early pregnancy on the infiltration by cells of the immune system in the sow endometrium. *Anim Reprod Sci*, 2003, 75, 55-71.

Kho, E. M., L. M. McCowan, et al. Duration of sexual relationship and its effect on preeclampsia and small for gestational age perinatal outcome. *J Reprod Immunol*, 2009, 82, 66-73.

Kimura, H., A. Fukui, et al. Timed sexual intercourse facilitates the recruitment of uterine CD56 (bright) natural killer cells in women with infertility. *Am J Reprod Immunol*, 2009, 62, 118-124.

Kirsch T. M., A. C. Friedman, R. L. Vogel, G. L. Flickinger. Macrophages in corpora lutea of mice: characterization and effects on steroid secretion. *Biol Reprod*, 1981, 25, 629-38.

Kruger T. H., B. Leeners, E. Naegeli, et al. Prolactin secretory rhythm in women: immediate and long-term alterations after sexual contact. *Hum Reprod*, 2012, 27, 1139-43

Langendijk, P., E. G. Bouwman, et al. Myometrial activity around estrus in sows: spontaneous activity and effects of estrogens, cloprostenol, seminal plasma and clenbuterol. *Theriogenology*, 2002, 57, 1563-1577.

Langendijk, P., E. G. Bouwman, et al. Effects of different sexual stimuli on oxytocin release, uterine activity and receptive behavior in estrous sows. *Theriogenology*, 2003, 59, 849-861.

Lattanzi M. L., C. B. Santos, M. D. Mudry, et al. Exposure of bovine oocytes to the endogenous metabolite 2-methoxyestradiol during in vitro maturation inhibits early embryonic development. *Biol Reprod*, 2003, 69, 1793-1800.

Lee, C., S. M. Sintich, et al. Transforming growth factor-beta in benign and malignant prostate. *Prostate*, 1999, 39, 285-290.

Lehrer, R. I., T. Ganz, et al. Neutrophils and host defense. *Ann Intern Med*, 1988, 109, 127-142.

Le Mellay V., B. Grosse, M. Lieberherr. Phospholipase C b and membrane action of calcitriol and estradiol. *J Biol Chem*, 1997, 272, 11902-11907.

Lösel R, M. Wheling. Nongenomic actions of steroid hormones. *Nature Reviews Molecular Cell Biology*, 2003, 4, 46-56.

Lovell, J. W. and R. Getty. Fate of semen in the uterus of the sow: histologic study of endometrium during the 27 hours after natural service. *Am J Vet Res*, 1968, 29, 609-625.

Matsuyama S., K. Shiota, M. Takahashi. Possible role of transforming growth factor-beta as a mediator of luteotropic action of prolactin in rat luteal cell cultures. *Endocrinology*, 1990, 127, 1561-7

McMaster, M. T., R. C. Newton, et al. Activation and distribution of inflammatory cells in the mouse uterus during the preimplantation period. *J Immunol*, 1992, 148, 1699-1705.

Maegawa, M., M. Kamada, et al. A repertoire of cytokines in human seminal plasma. *J Reprod Immunol*, 2002, 54, 33-42.

Mah, J., J. E. Tilton, et al. (1985). The effect of repeated mating at short intervals on reproductive performance of gilts. J Anim Sci **60**(4): 1052-1054.

Moldenhauer, L. M., K. R. Diener, et al. Cross-presentation of male seminal fluid antigens elicits T cell activation to initiate the female immune response to pregnancy. *J Immunol*, 2009, 182, 8080-8093.

Murray, F. A., A. P. Grifo, Jr., et al. Increased litter size in gilts by intrauterine infusion of seminal and sperm antigens before breeding. *J Anim Sci*, 1983, 56, 895-900.

Murray, F. A., A. P. Grifo, Jr., et al. Increased litter size in gilts by presensitization with boar semen. *J Anim Sci*, 1986, 196(Suppl), 189.

Nicholson, H. D. Oxytocin: a paracrine regulator of prostatic function. *Rev Reprod*, 1996, 69-72.

Niswender G. D., J. L. Juengel , P. J. Silva , M. K. Rollyson , E. W. McIntush Mechanisms controlling the function and life span of the corpus luteum. *Physiol Rev*, 2000, 80, 1-29

Nocera, M. and T. M. Chu. Transforming growth factor beta as an immunosuppressive protein in human seminal plasma. *Am J Reprod Immunol*, 1993, 30, 1-8.

O'Leary, S., S. A. Robertson, et al. The influence of seminal plasma on ovarian function in pigs--a novel inflammatory mechanism? *J Reprod Immunol*, 2002, 57, 225-238.

O'Leary, S., M. J. Jasper, et al. Seminal plasma regulates endometrial cytokine expression, leukocyte recruitment and embryo development in the pig. *Reproduction*, 2004, 128, 237-247.

O'Leary S, M. J. Jasper, G. M. Warnes, D. T. Armstrong, S. A. Robertson. Seminal plasma regulates endometrial cytokine expression, leukocyte recruitment and embryo development in the pig. *Reproduction*, 2004, 128, 237-47

O'Leary S, M. J. Jasper, S. A. Robertson, D. T. Armstrong. Seminal plasma regulates ovarian progesterone production, leukocyte recruitment and follicular cell responses in the pig. *Reproduction* 2006, 132, 147-58.

Orihuela P.A., H. B. Croxatto. Acceleration of oviductal transport of oocytes induced by estradiol in cycling rats is mediated by nongenomic stimulation of protein phosphorylation in the oviduct. *Biol Reprod* 2001, 65, 1238-1245.

Orihuela P. A., A. Parada-Bustamante, P. P. Cortes, et al. Estrogen receptor, cyclic adenosin monophosphate and protein kinase A are involved in the nongenomic pathway by which estradiol accelerates oviductal oocyte transport in cyclic rats *Biol Reprod* 2003, 68, 1225-1231.

Orihuela P. A., A. Parada-Bustamante, L. M. Zuñiga, et al. Inositol triphosphate participates in an oestradiol signalling pathway involved in accelerated oviductal transport in cycling rats. *J Endocrinol* 2006, 188, 579-588.

Orihuela P. A., L. M. Zuñiga, M. Rios, et al. Mating changes the subcellular distribution and the functionality of estrogen receptors in the rat oviduct. *Reprod. Biol. Endocrinol.*, 2009, 7, 139.

Orihuela P. A., M. Ríos, H. B. Croxatto. Disparate effects of estradiol on egg transport and oviductal protein synthesis in mated and cyclic rats. *Biol Reprod*, 2001, 65, 1232-1237.

Ormandy C. J., A. Camus, J. Barra, et al. Null mutation of the prolactin receptor gene produces multiple reproductive defects in the mouse. *Genes Dev*, 1997, 11, 167–78.

Oróstica M. L., L. M. Zuñiga, D. Utz, A. Parada-Bustamante, H. B. Croxatto, L. A. Velasquez, P. A. Orihuela. TNF-α y TGF-β signaling cascades associated with the changes in the mode of action of estradiol in the rat Oviduct. *Biol Reprod*, 2009; S:120-121.

Overstreet, J. W. and R. A. Tom. Experimental studies of rapid sperm transport in rabbits. *J Reprod Fertil*, 1982, **66**, 601-606.

Palm F, I. Walter, S. Budik, J. Kolodziejek, N. Nowotny, C. Aurich. Influence of different semen extenders and seminal plasma on PMN migration and on expression of IL-1beta, IL-6, TNF-alpha and COX-2 mRNA in the equine endometrium. *Theriogenology*, 2008, 70, 843-51.

Pandya, I. J. and J. Cohen. The leukocytic reaction of the human uterine cervix to spermatozoa. *Fertil Steril*, 1985, 43, 417-421.

Pate JL, K. Toyokawa, S. Walusimbi, E. Brzezicka. The interface of the immune and reproductive systems in the ovary: lessons learned from the corpus luteum of domestic animal models. *Am J Reprod Immunol*, 2010, 64, 275-86.

Parada-Bustamante A, P. A. Orihuela, M. Rios, et al. Catechol-O-Methyltransferase and Methoxyestradiols Participate in the Intraoviductal Nongenomic Pathway Through Which Estradiol Accelerates Egg Transport in Cycling Rats. *Biol Reprod*, 2007, 77: 934-941.

Parada-Bustamante A, P. A. Orihuela, M. Rios, et al. A non-genomic signaling pathway shut down by mating changes the estradiol-induced gene expression profile in the rat oviduct. *Reproduction*, 2010, 139, 631-644.

Parada-Bustamante A, P. A. Orihuela, H. B. Croxatto. Effect of intrauterine insemination with spermatozoa or foreign protein on the mechanism of action of estradiol in the rat oviduct. *Reproduction* 2003, 125, 677-682.

Parada-Bustamante A, P. A. Orihuela, H. B. Croxatto. Differential participation of endothelin receptors in the estradiol-induced egg transport acceleration in unmated and mated rats. *Asia Pac J Reprod* 2012, 1, 55-59.

Parada-Bustamante A, P. A. Orihuela. Sistema Inmune y mecanismo de acción del estradiol. XLV Reunión Anual de la Sociedad de Biología de Chile, 2002, Pucón-Chile.

Paria B. C., C. Chakraborty, S. K. Dey. Catechol estrogen formation in the mouse uterus and its role in implantation. *Mol Cell Endocrinol*, 1990, 69, 25–32.

Peñarroja-Matutano C, A. Parada-Bustamante, P. A. Orihuela, M. Rios, L. A. Velasquez, H. B. Croxatto. Genital Sensory Stimulation Shifts Estradiol Intraoviductal Signaling from Nongenomic to Genomic Pathways, Independently from Prolactin Surges. *Biol Res*, 2007, 40, 213-222.

Peters, B., T. Whittall, et al. Effect of heterosexual intercourse on mucosal alloimmunisation and resistance to HIV-1 infection. *Lancet*, 2004, 363, 518-524.

Pitjkanen, I. G. and M. I. Prokofjev. Modification of the sexual processes of sows by neurotropic substances and stimulation of the receptors of the uterine cervix. *Anim Breed*, 1964 Abstr 35, 124.

Portus B. J., T. Reilas, T. Katila. Effect of seminal plasma on uterine inflammation, contractility and pregnancy rates in mares. *Equine Vet J*, 2005, 37, 515-9

Portus, B. J., T. Reilas, et al. Effect of seminal plasma on uterine inflammation, contractility and pregnancy rates in mares. *Equine Vet*, 2005, J 37, 515-519.

Qui J, M. A. Bosch, S. C. Tobias, et al. Rapid signaling of estrogen in hypothalamic neurons involves a novel G-protein-coupled estrogen receptor that activates protein kinase C. *J Neurosci*, 2003, 23, 9529-9540.

Ratto M. H., W. Huanca, J. Singh, G. P. Adams. Local versus systemic effect of ovulation-inducing factor in the seminal plasma of alpacas. *Reprod Biol Endocrinol*, 2005, 3, 29.

Ratto M. H., L. T. Delbaere, Y. A. Leduc, R. A. Pierson, G. P. Adams. Biochemical isolation and purification of ovulation-inducing factor (OIF) in seminal plasma of llamas. *Reprod Biol Endocrinol*, 2011, 9, 24.

Tremellen K. P., R. F. Seamark, S. A. Robertson. Seminal transforming growth factor beta1 stimulates granulocyte-macrophage colony-stimulating factor production and inflammatory cell recruitment in the murine uterus. *Biol Reprod*, 1998, 58, 1217-25.

Reuquén P., M. L. Orostica, P. A. Orihuela. Effect of the sensory stimulation on cellular signaling associated to estradiol in the rat pituitary. *15th Congress of the European Neuroendocrine Association*, 12-15 September 2012, Austria-Vienna

Rhodes, M., J. H. Brendemuhl, et al. Litter characteristics of gilts artificially inseminated with transforming growth factor-beta. *Am J Reprod Immunol*, 2006, **56**, 153-156.

Rios M, A. Parada-Bustamante, L. A. Velasquez, et al. A s100 calcium binding protein G is involved in the Genomic Pathway of estradiol that accelerates oviductal egg transport in mated rats. *Reprod Biol Endocrinol*, 2011; 9:69

Robertson S. A. Seminal plasma and male factor signalling in the female reproductive tract. *Cell Tissue Res*, 2005, 322:43-52.

Robertson S. A. Seminal fluid signaling in the female reproductive tract: lessons from rodents and pigs. *J Anim Sci*, 2007, 85, E36-44.

Robertson, S. A. Seminal plasma and male factor signalling in the female reproductive tract. *Cell Tissue Res*, 2005, 322, 43-52.

Robertson, S. A., L. R. Guerin, et al. Activating T regulatory cells for tolerance in early pregnancy the contribution of seminal fluid. *J Reprod Immunol*, 2009, 83, 109-116.

Robertson, S. A., W. V. Ingman, et al. Transforming growth factor beta--a mediator of immune deviation in seminal plasma. *J Reprod Immunol*, 2002, 57, 109-128.

Rozeboom, K. J., G. Rocha-Chavez, et al. Inhibition of neutrophil chemotaxis by pig seminal plasma in vitro: a potential method for modulating post-breeding inflammation in sows. *Reproduction*, 2001, 121, 567-572.

Robertson, S. A., V. J. Mau, et al. Role of high molecular weight seminal vesicle proteins in eliciting the uterine inflammatory response to semen in mice. *J Reprod Fertil*, 1996, 107, 265-277.

Rozeboom, K. J., M. H. Troedsson, et al. Characterization of uterine leukocyte infiltration in gilts after artificial insemination. *J Reprod Fertil*, 1998, 114, 195-199.

Rozeboom, K. J., M. H. Troedsson, et al. The effect of spermatozoa and seminal plasma on leukocyte migration into the uterus of gilts. *J Anim Sci*, 1999, 77, 2201-2206.

Sanford, T. R., M. De, et al. Expression of colony-stimulating factors and inflammatory cytokines in the uterus of CD1 mice during days 1 to 3 of pregnancy. *J Reprod Fertil*, 1992, 94, 213-220.

Scott, J. L., N. Ketheesan, et al. Leucocyte population changes in the reproductive tract of the ewe in response to insemination. *Reprod Fertil Dev*, 2006, 18, 627-634.

Sharkey, D. J., K. P. Tremellen, et al. Seminal fluid induces leukocyte recruitment and cytokine and chemokine mRNA expression in the human cervix after coitus. *J Immunol*, 2012, 188, 2445-2454.

Sharkey D. J., A. M. Mcpherson, K. P. Tremellen, S. A. Robertson. Seminal plasma differentially regulates inflammatory cytokine gene expression in human cervical and vaginal epithelial cells. *Mol Hum Reprod*, 2007, 13, 491-501.

Silverman, A. J. and E. A. Zimmerman. Magnocellular neurosecretory system. *Annu Rev Neurosci*, 1983, 6, 357-380.

Singh S., P. D. Gupta. Induction of phosphoinositide-mediated signal pathway by 17b-oestradiol in rat vaginal epithelial cells. *J Mol Endo*, 1997, 19, 249-257.

Smith M. S., M. E. Freeman, J. D. Neill. The control of progesterone secretion during the estrous cycle and early pseudopregnancy in the rat: prolactin, gonadotropin and steroid levels associated with rescue of the corpus luteum of pseudopregnancy. *Endocrinology*, 1975, 96, 219-26.

Sobrero, A. J. and J. Macleod. The immediate postcoital test. Fertil Steril, 1962, 13, 184-189.

Suarez S.S., A. A. Pacey. Sperm transport in the female reproductive tract. *Hum Reprod Update*, 2006, 12, 23-37.

Tanco V. M., M. D. Van Steelandt, M. H. Ratto, G. P. Adams. Effect of purified llama ovulation-inducing factor (OIF) on ovarian function in cattle. *Theriogenology*, 2012 (In press).

Taneike, T., H. Miyazaki, et al. Autonomic innervation of the circular and longitudinal layers in swine myometrium. *Biol Reprod*, 1991, 45, 831-840.

Thomas P. G., G. G. Ignotz, B. A. Ball, S. P. Brinsko, W. B. Currie. Effect of coculture with stallion spermatozoa on de novo protein synthesis and secretion by equine oviduct epithelial cells. *Am J Vet Res*, 1995, 56, 1657-62

Thompson, L. A., C. L. Barratt, et al. The leukocytic reaction of the human uterine cervix. *Am J Reprod Immunol*, 1992, 28, 85-89.

Toner, J. P. and N. T. Adler. Influence of mating and vaginocervical stimulation on rat uterine activity. *J Reprod Fertil*, 1986, 78, 239-249.

Toner, J. P. and N. T. Adler. The role of uterine luminal fluid in uterine contractions, sperm transport and fertility of rats. *J Reprod Fertil*, 1985, 74, 295-302.

Tremellen, K. P., D. Valbuena, et al. The effect of intercourse on pregnancy rates during assisted human reproduction. *Hum Reprod*, 2000, 15, 2653-2658.

Tremellen, K. P., R. F. Seamark, et al. Seminal transforming growth factor beta1 stimulates granulocyte-macrophage colony-stimulating factor production and inflammatory cell recruitment in the murine uterus. *Biol Reprod*, 1988, 58,1217-1225.

Troedsson, M., C. Concha, et al. A preliminary study of uterine derived polymorphonuclear cell function in mares with chronic uterine infections. *Acta Vet Scand*, 1990, 31, 187-192.

Troedsson, M. H., I. K. Liu, et al. Multiple site electromyography recordings of uterine activity following an intrauterine bacterial challenge in mares susceptible and resistant to chronic uterine infection. *J Reprod Fertil*, 1993, 99, 307-313.

Vandemark, N. L. and R. L. Hays. Rapid sperm transport in the cow. *Fertil Steril*, 1954, 5, 131-137.

Ventura, W. P. and M. Freund. Evidence for a new class of uterine stimulants in rat semen and male accessory gland secretions. *J Reprod Fertil*, 1973, 33, 507-511.

von Euler, U. S. On the specific vaso-dilating and plain muscle stimulatory substances from accessory genital glands in man and certain animals, (prostaglandin and vesiglandin). J. *Physiol., Lond.,* 1936, 88: 213.

Waberski D., A. Döhring, F. Ardón, N. Ritter, H. Zerbe, H. J. Schuberth, M. Hewicker-Trautwein, K. F. Weitze, R. H. Hunter. Physiological routes from intra-uterine seminal contents to advancement of ovulation. *Acta Vet Scand*, 2006, 48:13.

Watanabe K, K. Takanashi, S. Imaoka, et al. Comparison of cytochrome P-450 species which catalyze the hydroxylations of the aromatic ring of estradiol and estradiol 17-sulfate. *J Steroid Biochem Biol*, 1991, 38, 737–743.

Watson, J. G., J. Carroll, et al. Reproduction in mice: The fate of spermatozoa not involved in fertilization. *Gamete Res*, 1983, 7, 75-84.

Welboren W.J., F. C. Sweep, P. N. Span, et al. Genomic actions of estrogen receptor alpha: what are the targets and how are they regulated?. *Endocr Relat Cancer*, 2009, 16, 1073-89.

Yamanouchi K., S. Matsuyama, M. Nishihara, et al. Takahashi. Splenic macrophages enhance prolactin-induced progestin secretion from mature rat granulosa cells in vitro. *Biol Reprod*, 1992, 46, 1109-13.

In: Human and Animal Mating
Editors: M. Nakamura and T. Ito

ISBN: 978-1-62417-085-0
© 2013 Nova Science Publishers, Inc.

Chapter 2

SEXUAL MATURATION, MATING STRATEGIES AND NEUROENDOCRINOLOGY IN SOCIAL INSECTS

Ken-ichi Harano[1] and Ken Sasaki[2]
[1]Honeybee Science Research Center, Tamagawa University,
Machida, Tokyo, Japan
[2]Department of Applied Bioscience, Human Information Systems,
Kanazawa Institute of Technology, Hakusan, Ishikawa, Japan

ABSTRACT

The process of mating in social insects is unique among insect species because reproductive individuals interact with other colony members including reproductives and non-reproductives. In some species, reproductives go through intra-caste competition to monopolize the reproduction in a colony before mating. They show diverse strategies for this competition. Reproductives may undergo some physiological and behavioral changes prior to the onset of mating flight as well as dynamic changes after mating. Although the mechanisms underlying these changes differ depending on the species and sex, particular neuroendocrines appear to regulate the changes in some species. In honeybees, dopamine is likely to be involved in behavioral activation before mating. Physiological factors affecting the amount of brain dopamine may differ between the sexes. Sexual difference in dopamine regulation might result from their different social roles in the colony. The present chapter also addresses multiple mating as a queen's mating strategy. Some queens are monoandrous and mate only once, but others are polyandrous and may continue mating attempts until they mate with a sufficient number of males. Honeybee queens are likely to monitor the total amount of semen received in the previous mating to determine the timing of cessation of mating flight. By multiple mating, they increase the genetic heterogeneity in workers and enhance colony performance. Multiple mating also achieves lower intra-colony relatedness and subsequently reduces the queen-worker conflicts over sexual production.

1. INTRODUCTION

Social insects form a kin-based group called a colony, constructing an orderly society. Some species with highly evolved sociality (eusociality) exhibit a reproductive skew or a division of reproduction among individuals. In such insects, individuals express specialized morphology and behavior for their social roles, thereby achieving an efficient division of labor referred to as a caste; females of the reproductive caste are called queens and those of the non-reproductive caste workers.

In social Hymenoptera, including bees, wasps and ants, queens and workers develop from female eggs, expressing different phenotypes from the same genome depending on the environment and individual nutrition; this is a phenomenon called polyphenism, seen in insect species other than eusocial insects, such as phase changes in locusts (Pener and Simpson, 2009; Verlinden et al., 2009) and seasonal changes in butterflies (Shapiro, 1976; Brakefield, 1996).

Social insects produce reproductive individuals (queens and males) before and during the reproductive season. In social Hymenoptera, females develop from fertilized eggs, whereas males develop from unfertilized eggs parthenogenetically. This sex-determination system enables queens to control the sex of a brood in terms of whether or not they deposit sperm on the eggs at oviposition, and workers are obliged to work within this constraint. The queens, therefore, have the ability to determine when to produce males or non-reproductive females, workers, for colony maintenance.

It has been reported that queens determine the initial sex ratio in response to environmental factors such as availability of food, temperature and the degree of colony growth (Gösswald and Bier, 1955; Hölldobler and Wilson, 1990; Aron et al., 1994; Sasaki et al., 1996; Sasaki and Obara, 2001). Workers can distinguish the sex of larvae, provide preferential care for either sex or eliminate male larvae (Aron et al., 1995; Keller et al., 1996; Passera and Aron, 1996; Chapuisat et al., 1997; Sasaki et al., 2004). Workers can also determine the female caste by manipulating the type of food provided to female larvae (Michener, 1974; Hölldobler and Wilson, 1990; Kamakura, 2011). Thus, workers determine the allocation of investment in sexual production and worker production for colony growth and maintenance.

Reproductive individuals in highly eusocial species show limited behavioral repertoires for reproduction, in contrast to various behavioral repertoires in workers. Virgin queens usually emerge at the same time as male emergence, reach sexual maturity in the nest and make mating flights. After mating, their ovaries develop intensely and they start laying eggs and never fly to mate again once egg laying has started (Ruttner, 1956).

These changes are probably adaptive transitions from mating-oriented to laying-oriented queens because different activities are required in virgin and mated queens (Julian and Gronenberg, 2002; Burns et al., 2007). Mated queens in eusocial species should be specialized for mass egg production, but some traits for egg production, for example, developed ovaries, may be a burden for activities of virgin queens like mating flight.

Adult males exhibit behaviors specialized for reproduction, and do not display a division of labor in a highly eusocial Hymenoptera (Winston, 1987). Behaviors of sexually mature males mainly involve flying in order to mate with a queen.

However, until the males are sexually mature, they are present in the nest and fed by workers. Males in several highly eusocial bees including honeybees mate only once and die after mating by their genitalia being torn off, while others in primitively and highly eusocial species may be able to mate several times and die a natural death.

Sexual behaviors in social insects may be regulated by particular neuroendocrine mechanisms and influenced by the maturation of reproductive organs. Mating activities in both sexes in several species can be controlled by biogenic amines. Biogenic amines play physiological roles as neurotransmitters, neuromodulators and neurohormones in insects. Some of the biogenic amines promote the development of reproductive organs and activate reproductive behaviors.

In the present chapter, we introduce the sexual behaviors of both queens and males and the physiological regulation systems in social insects. Comparative study of the physiological regulation systems among species with different levels of social evolution should help us to understand the evolution of the division of reproduction in social insects.

2. BEHAVIOR BEFORE MATING IN REPRODUCTIVES

(1) Queens

Social insects can be classified into 2 groups with respect to the presence or absence of worker cooperation for starting a new colony. In bumble bees, social wasps, and many species of ants and termites, queens start a colony alone or in cooperation with other reproductives, but without workers. Those queens leave the nest when sexually mature and hibernate or start founding a colony after mating, but do not return to the mother nest. Other social insects including honeybees, stingless bees and some ants multiply their colony by budding or swarming, in which queens leave the mother colony with workers to establish a new colony. In these species, virgin queens temporarily leave the colony for a short time for mating, usually by flying. Before leaving the nest, they learn the location of their nest for returning. In honeybee queens, flight begins at day 6 – 9 after emergence and the first few flights are regarded as orientation flights during which they learn geographical features around the nest. After one or a few orientation flights, honeybee queens make 1 – 5 mating flights over several days. They fly as far as several km to a mating site, which is called a drone congregation area (DCA), located 10 – 40 m above the ground, and mate with multiple males that come from nearby colonies.

Queens return to the colony relying on their memory, but workers may also assist them in their return. When honeybee queens travel away for mating, some workers move to the nest entrance and perform fanning while exposing their Nasonov gland in order to disperse a pheromone (Ruttner, 1956). This pheromone may help queens to locate their nest on the return trip after mating.

Several aspects of behavioral changes are known in queens before mating. For example, honeybee virgin queens become photopositive upon departing for mating flight, whereas they are negatively phototactic for several days after emergence (Berthold and Benton, 1970). This change appears to be reasonable given that, for mating, they leave the dark nest for the brightness outside. When spontaneous locomotor activity was quantified for isolated virgin queens in small circuits, it was found to increase with age after emergence and reached a peak at the age at which queens make a mating flight (Harano et al., 2007). This observation suggests that queens are behaviorally activated before mating.

Some new queens may be eliminated before mating in species that multiply the colony by colony fission. When a colony produces more queens than necessary, excess new queens are killed by workers in the cases of fire ants (Fletcher, 1986) and stingless bees (Michener, 1974). However, in honeybees, mutual aggression among new queens is the major process for elimination of excess queens. Surviving this competition is essential for virgin queens to mate because mating is conducted after all virgin queens except one are eliminated.

Honeybees raise new queens in queen cells, which are special chambers for queen rearing, built during the reproductive season. When new queens grow to pupae in queen cells, the old queens (the mothers of the new queens) leave the nest with a primary swarm consisting of approximately half of the colony workers in order to establish a new colony. The rest of the workers, broods and comb with honey and pollen will be taken over by a new queen. If a sufficient number of workers remain after a primary swarm, some new queens may leave the colony by forming secondary swarms. Otherwise, all virgin queens are involved in lethal competition for succession in the colony.

New queens eliminate sister queens in two ways: "assassination" and "duel". When they encounter an immature sister residing in a queen cell, they make a hole in the queen cell wall with their mandibles and kill the sister inside (referred to as "assassination" or queen cell destruction). It is more difficult and risky for a virgin queen to eliminate adult sisters that have emerged from queen cells. Adult queens recognize one another using a contact cue (Pflugfelder and Koeniger, 2003) and kill rivals by stinging ("duel"). Although older virgin queens are more likely to win the duel than younger ones (Tarpy et al., 2000), each participant in the duel has a risk of being killed or injured.

Virgin queens show diverse strategies for this competition (Bernasconi et al., 2000; Gilley, 2001; Tarpy and Fletcher, 2003; Gilley and Tarpy, 2005). For example, they identify queen cells housing sisters close to emergence using olfactory and vibratory/acoustic cues (Harano and Obara, 2004a) and destroy them earlier than those housing younger sisters (Caron and Greve, 1979; Harano et al., 2004b). Because destroying queen cells requires more than 1 h in some cases (Harano and Obara, 2008a), the selective elimination of mature sisters is likely to reduce the number of emerged rivals and subsequent risky duels. After making a hole in a queen cell wall, virgin queens sting the sister inside the cell by inserting their abdomen through the hole when the sister has completed adult eclosion (Fletcher 1978; Harano et al., 2008a). For a pupal sister, virgin queens leave it without stinging after queen cell destruction, and workers instead pull the pupa out from the cell and eventually kill it.

Virgin honeybee queens produce two types of sound in the colony, called tooting and quacking signals, by contracting their flight muscles. The former is produced by emerged queens and the latter by pre-emerging queens residing in queen cells. This behavior is called queen piping and is characteristic of virgin queens. Interestingly, pre-emerging queens often perform quacking in response to tooting made by an emerged queen (Michelsen et al., 1986). It has also been reported that the tooting signal stimulates workers to confine pre-emerging queens to their queen cells and causes a delay in their emergence (Grooters, 1987). These observations suggest that these sounds serve as a means of communication or as a strategy for competition among virgin queens.

Some observations suggest that workers indirectly influence the outcome of competition by various means, such as agonistic interactions and mechanical stimulation of particular queens (Gilley, 2001; Schneider et al., 2001), although they do not apparently intervene in duels of queens. It has also been reported that the relatedness of queens to surrounding workers affects the outcome of duels (Tarpy and Fletcher, 1998).

(2) Males

Males of social insects show various mating behaviors depending on the mating system of their species. In termites, males form mating pairs on the ground after swarming and start a colony with a new queen. Termite queens have a small spermatheca and cannot keep a large amount of sperm in their body. Instead, their males live as long as queens and repeat mating regularly. Males of most eusocial Hymenoptera, in contrast, have a short life-span and die after one or a few matings. The local density and degree of dispersal of new queens are suggested to be two key factors determining the mating system and males' mating behavior in social insects (Boomsma et al., 2005). In bumble bees and social wasps in which new queens emerge asynchronously and live alone outside of the mother nest before mating, males patrol the female's foraging range or flight corridors. In many ants, termites, some social bees and wasps, the emergence of new queens occurs in a synchronized fashion to some extent. Their new queens and males typically aggregate at a landmark and mating is conducted in a mating swarm. Mating also occurs in the nest or at the colony entrance in some species (Akre, 1982; Kinomura and Yamauchi, 1987).

The behavioral development of males is studied best in honeybees. They are fed entirely by workers for a few days after emergence and gradually start to eat honey from cells by themselves (Free, 1957). Males have a large amount of honey, as much as 20 mg, in the honey stomach upon departure for mating and most of this honey is consumed during the mating flight (Free, 1957). The location at which they reside within the nest changes with age. Young males tend to stay in the brood-rearing area located in the center of nest, whereas old males are found on honeycombs and the floor of hive, in peripheral areas (Free, 1957). They make their first flight at day 7 - 9 after emergence (Ruttner, 1966) and perform a mating flight to the DCA following several orientation flights, typically later than day 12. Phototactic change with age is also known in males, as in queens. They become photopositive rather than photonegative before performing flights (Berthold and Benton, 1970).

In the DCA, males locate new queens in the air using visual and pheromonal cues (Gary, 1962, 1963; Gary and Marston, 1971; Praagh et al., 1980; Menzel et al., 1991). When they mount a flying queen, their endophallus is explosively everted into the sting chamber of the queen and the ejaculate is transferred to the queen's genital tract (Winston, 1987). Males become paralyzed at this moment and drop to the ground a few seconds later by tearing their genitalia, which is left in the queens. These males die shortly after dropping. When they do not mate in the DCA, they return to the colony and make another mating flight on the same or another day.

Unlike queens, honeybee males do not actively interact with one another in the colony. This is true for the majority of social insects. In some ants, however, lethal fights among males occur before mating (Kinomura and Yamauchi, 1987; Heinze et al., 1998). Several species of *Cardiocondyla* ants produce two types of male: alates and ergatoids. Alate males are winged and conduct mating flights, although they mate within the natal nest when virgin females are present (Kinomura and Yamauchi, 1987; Yoshizawa et al., 2011). Ergatoid males are wingless and carry out only intra-nest mating. This type of male fights other males and kills rivals using its long sickle-shaped mandibles to monopolize females in the nest (Kinomura and Yamauchi, 1987). It also kills eclosing pupae of ergatoids, as occurs in competition in honeybee queens. Unlike in honeybees, workers may also attack one of two ergatoids that are fighting and cause much damage to this male.

Ergatoids of *Cardiocondyla* are also aggressive to alate males. Their aggression would be potentially lethal only for young alates because aged alates can move quickly and escape them. It has been reported that young alates mimic the chemical profile of a virgin queen's cuticular hydrocarbon to avoid the aggression of ergatoids (Cremer et al., 2002). Mature alates show a distinct profile of cuticular hydrocarbon from that of virgin queens, probably because the chemical cue may be used for mate recognition by females in nuptial-flight mating.

3. SEXUAL MATURATION AND ITS PHYSIOLOGICAL MECHANISMS

(1) Queens

Queens of social insects are specialized for reproduction in terms of their internal and external morphology, physiology and behavior. For example, honeybee queens have a pair of large ovaries consisting of 160 - 180 ovarioles (Snodgrass, 1956) and are capable of ovipositing more than 2000 eggs a day. Although this ability is exerted only after ovarian development triggered by mating, queens proceed initial processes of oogenesis before mating (Tanaka and Hartfelder, 2004). Their ovaries are filled with undifferentiated cells at emergence and these cells differentiate into oocytes and nurse cells by day 8 in the basal part of the ovarioles (Patricio and Cruz-Landim, 2002). When sexually-matured queens are prevented from mating, programmed cell death and cell reabsorption may occur in the ovarioles (Patricio and Cruz-landim, 2002; Tanaka and Hartfelder, 2004).

The physiological mechanism allowing oocytes to accumulate yolk after mating differs depending on the species and degree of sociality. In many solitary insects, juvenile hormone (JH) plays an important role in the regulation of ovarian development as well as that of reproductive behavior (Wyatt and Davey, 1996). This hormone also acts as a key regulator of reproduction in the primary eusocial insects, such as bumble bees (Bloch et al., 1996, 2000a) and paper wasps (Bohm, 1972; Roseler et al., 1980). However, this is not the case in honeybees (Robinson et al., 1992). JH, instead, gains a new function to regulate age-based polyethism of workers in this insect (Robinson, 1985; Sasagawa et al., 1989; Robinson and Vargo, 1997; Hartfelder, 2000). Although the function of JH is still unknown in honeybee queens, it does not appear to regulate their mating flight and ovarian development because its hemolymph titer does not show obvious association with such events (Fahrbach et al., 1995).

Dopamine is another candidate for a regulator of ovarian development in honeybees. Biogenic amines including dopamine are involved in the regulation of various physiological and behavioral features in a wide range of animals (Evans, 1980; Carlson, 2003; Sasaki and Harano, 2010). In solitary (Neckameyer, 1996; Pendleton et al., 1996; Gruntenko et al., 2005) and social insects (Bloch et al., 2000b; Boulay et al., 2001; Sasaki et al., 2007, 2009), dopamine is suggested to have gonadotropic functions. This substance is also demonstrated to promote ovarian development under queenless conditions in honeybee workers (Harris and Woodring, 1995; Sasaki and Nagao, 2001; Dombroski et al., 2003). The fact that the brain level of dopamine is higher in queens than in workers (Brandes et al., 1990; Harano et al., 2005; Sasaki et al., 2012) implies its reproductive function in this species. However, the relationship between dopamine levels and ovarian status is puzzling in queens: their brain and hemolymph dopamine levels are constantly high from emergence until mating in virgin queens with undeveloped ovaries and they decline greatly within a few days after mating, which triggers ovarian development (Harano et al., 2005, 2008b). Thus, it has not been concluded whether this substance has a gonadotropic function in honeybee queens.

Dopamine may play an important role in behavioral activation in virgin queens. This substance stimulates locomotor activity in various animals including vertebrates (Beninger, 1983; Carlson, 2003) and invertebrates (Sasaki and Harano, 2010). As mentioned above, both dopamine and locomotor activity levels are high in sexually mature virgin queens in honeybees. When a dopamine-receptor antagonist is injected into virgin queens, locomotor activity is depressed (Harano et al., 2008b), suggesting that they are behaviorally activated by dopamine. In workers, high dopamine levels are also associated with extensive motor activities. Foragers required to fly long distances show higher brain dopamine levels than those engaged in other tasks such as cleaning and nursing (Wagener-Hulme et al., 1999).

Behavioral activation may be particularly important for honeybee queens because they go through not only mating but also intra-caste competition. A small difference in physical capacity could greatly influence the outcome of such competition. An activated motor system is also required for mating, during which queens fly several km from the nest to the DCA. After mating, honeybee queens exhibit decreased dopamine levels and locomotor activity (Harano et al., 2005; 2007; 2008b). Dopamine is likely to be involved in the behavioral activation for these activities.

It has recently been suggested that the large amounts of dopamine found in reproductive individuals result from intensive provisioning from other colony members (Sasaki et al., 2012). When activities of dopa decarboxylase, an enzyme that synthesizes dopamine from dopa, were compared between virgin honeybee queens and their workers at the same age, no statistical difference was detected. The queens have a larger amount of dopa in the brain than workers, and oral application of dopa increased the brain levels of dopa and dopamine in workers, suggesting that the caste difference in dopamine levels results from the difference in the amount of its precursor in the brain. The large amount of dopa is assumed to be derived from royal jelly fed by workers. Royal jelly is a mixture of secretions from the hypopharyngeal and mandibular glands of young workers and contains a large amount of various amino acids including tyrosine, the precursor of dopa (Townsend and Lucas, 1940; Haydak, 1970). This special food is mainly fed to larval and adult queens. If dopamine plays an important role in competition and mating by activating the motor system, workers could influence such behaviors through its provisioning.

(2) Males

In social Hymenoptera, spermatogenesis is usually completed by adult emergence or at an early stage of adult life (Kerr and Silveira, 1974), although males of some ants continuously produce sperm throughout their entire adult life (Heinze et al., 1998). The produced sperm is transferred from testes to the seminal vesicles. The sperm transfer is almost completed by day 8 – 9 after emergence when the males begin orientation flights in honeybees (Jaycox, 1961). Coincident with the sperm transfer, their testes shrink and lose weight. The accessory reproductive glands are filled with mucus during this period (Moors et al., 2005). The mucus is ejaculated with sperm at copulation and forms a part of the "mating sign" left in the queen's sting chamber (see below).

The role of ecdysteroids in the development of male reproductive organs has been investigated in honeybees. Their hemolymph ecdysteroid titer shows a peak during the early pupal period (Tozetto et al., 2007) and sharply decreases on the first day of adult life (Colonello and Hartfelder, 2003). Injecting 20-hydroxyecdysone into newly emerged males is demonstrated to interfere the development of accessory gland (Colonello and Hartfelder, 2003). Ecdysteroids may negatively regulate the sexual maturation in males.

JH promotes flight in honeybee males. The synthesis and hemolymph levels of this hormone peak at ages when males start flying, and topical application of methoprene, a JH analog, or JH III to newly emerged males accelerates the onset of flight (Tozetto et al., 1995, 1997; Giray and Robinson, 1996). As no association between JH hemolymph titer and the occurrence of flight is found in queens (Fahrbach et al., 1995), the mechanism regulating mating behavior must differ between queens and males. Interestingly, JH promotes foraging flight in workers, suggesting that the flight is regulated by a common endocrinological mechanism in males and workers (Giray and Robinson, 1996).

JH controls brain dopamine levels in honeybee males, whereas no linkage between JH and dopamine has been reported in their queens and workers. The dopamine levels change linearly with their hemolymph JH titer, and application of methoprene induces an increase of dopamine in the brain (Harano et al., 2008c).

The expression of a dopamine receptor was also found to be up-regulated by methoprene treatment (Sasaki et al., in press). Because dopamine enhances locomotor activity in males (Akasaka et al., 2010), as found in queens and workers, this substance is suggested to mediate behavioral activation for mating under the control of JH in males.

Unlike in honeybees, both JH and dopamine have gonadotropic function in the females of bumble bees and social wasps (Bohm et al., 1972; Roseler et al., 1980; Bloch et al., 2000ab; Sasaki et al., 2007, 2009), as well as in some ants. Although the relationship between JH and dopamine has not been investigated yet in these species, it is possible that JH regulates reproduction via dopamine in the primitively eusocial insects. During social evolution, the linkage between JH and dopamine might have been lost because JH gained a new function for age-polyethism in workers, whereas it is retained in males that do not show age-polyethism.

4. CHANGES AFTER MATING

(1) Queens

Mating triggers dynamic changes in various aspects of physiology and behavior in queens (Table 1; Julian and Gronenberg, 2002). Some changes in mated females may be caused by bioactive substances transferred from males. Substances affecting females' mating acceptability, oogenesis, ovulation and so on are found in some solitary insects (Chapman, 2001; Gillott, 2003;

Yamane et al., 2008ab; Yamane and Miyatake, 2010). A bioactive substance influencing female behavior is also found in a bumble bee, *Bumbus terrestris.*

Table 1. Changes in physiology, behavior and gene expression after mating in *Apis mellifera* queens

Traits	Virgin	Mated	References
Ovaries	Undeveloped	Developed	
Vitellogenin titer in hemolymph	Low	High	Engels et al., 1990
QMP synthesis and release	Low	High	Engels et al., 1997; Richard et al., 2007; Kocher et al., 2009
Amount of amino acids in hemolymph	Small	Large	Hrassingg et al., 2003
Neuropil ratio in mushroom body	Small	Large	Fahrbach et al., 1995
Antennal lobe	Small	Large	Arnold et al., 1988
Dopamine levels	High	Low	Harano et al., 2005, 2008b
Mating flight	Present	Absent	
Oviposition	Absent	Present	
Cell inspection behavior	Rare	Frequent	Ohtani, 1985
Circadian rhythmicity in behavior	Present?	Absent	Free et al., 1992; Harano et al., 2007; Johnson et al., 2010
Phototaxis	Negative or Positive	Negative	Berthold and Benton, 1970
Frequency of receiving food from workers	Low	High	Ohtani, 1985
Piping	Present	Absent	
Expression of *Amfor*	High	Low	Richard et al., 2007
Expression of *Amdat*	High	Low	Nomura et al., 2009

See also Kocher et al. (2008) for changes in gene expression.

Males of this species form a mating plug wtih accessory gland secretion in the female genital tract, and linoleic acid contained in the secretion greatly decreases the female's mating acceptability (Bear et al., 2001). However, such effects are not known in the accessory gland secretion of honeybee males. Colonello and Hartfelder (2005) reported that injecting extracts of male mucus into the genital cavity of virgin queens at various concentrations did not affect their oogenesis. The post-mating changes are probably triggered by another mechanism in this insect. It is more likely that mechanical stimuli perceived by female organs at mating induce the changes, as described below.

The post-mating changes could be regarded as a transition from an adaptive form for mating to one for laying. For example, queens develop ovaries after mating, probably because developed ovaries would interfere with mating and related activities. In honeybees, indeed, laying queens with developed ovaries are almost unable to fly whereas virgin queens can fly a long distance.

However, it is also possible that the physiological state constrains their flight capability. Mated queens exhibit decreased levels of dopamine, which may be related to behavioral activation, as mentioned above. They could not fly because of an inactivated motor system. Since mated queens fly again upon reproductive or absconding swarming in honeybees, it would be interesting to investigate whether they have similar physiological states to virgin queens.

(2) Males

In some social insects including honeybees, males can mate only once because they die shortly after mating. In other social insects such as *Cardiocondyla* ants, males repeatedly mate (Kinomura and Yamauchi, 1987; Heinze et al., 1998), but changes in their behavior are not known, unlike in queens.

5. MATING STRATEGY ON COLONY GROWTH AND REPRODUCTIVE SUCCESS IN SOCIAL INSECTS

A number of mating events with different males and the utilization of sperm for egg fertilization by a queen influence the average genetic relatedness within a colony. Queens may be able to increase the number of mating events and mix spermatozoa from different males. These actions may be part of a queen's reproductive strategy because a multiple-mated queen can maintain a lower average relatedness in a colony than a single-mated queen, which would decrease queen-worker conflict over investments between new queens and males or between reproduction and colony maintenance (Crozier and Pamilo, 1996). A queen's multiple mating may also be adaptive for storage of a sufficient volume of semen in the spermatheca (Cole, 1983).

Multiple mating by queens is well known in honeybees. Honeybee queens mate with 12 males on average (Tarpy et al., 2004) and mix the spermatozoa in the spermatheca (Taber, 1955; Laidlaw and Page, 1984; Page and Metcalf, 1982). This results in genetic variations among workers in a colony (Estoup et al., 1994; Sasaki et al., 1995). Workers from different paternal lines may enhance the performance of the colony; for example, some workers from a particular paternal line can specialize as pollen foragers, while others from a different paternal line may prefer nectar (Page and Robinson, 1991). Thus, the queen's reproductive strategy to reduce conflicts between castes by multiple mating may also result in variations in worker behavior and successful colony growth and reproduction in the colony.

In several ant species, colonies with a multiple-mated queen produce only males for reproduction, whereas colonies with a single-mated queen produce only female reproductive individuals (Sundström, 1994). Since either single-mated or multiple-mated queens lay both female and male eggs, workers may manipulate the sex ratio of reproductive individuals depending on the average relatedness within a colony (Sundström et al. 1996). In this case, the number of matings by a queen seems to influence the colony character in terms of its production of reproductive individuals. However, it is uncertain whether or not this multiple mating decreases queen-worker conflict over investments between new queens and males.

The multiple mating strategy in a queen may originate from the morphology of the large spermatheca to store spermatozoa from several males. Honeybee queens have an extremely large spermatheca with a developing spermathecal pump system (Snodgrass, 1956). This spermatheca can store a large volume of semen, inactivate the spermatozoa for several years and discharge several spermatozoa and activate them for egg fertilization. If semen from males is not sufficient to expand the spermathecal wall mechanically, queens would continue to fly for mating again. It has been reported that storage of an insufficient volume of semen upon the first mating flight causes a second or third mating flight (Woyke, 1962; Kocher et al., 2008). Thus, the internal state of the spermatheca may provide feedback on the mating behaviors in queens. A large spermatheca may cause multiple mating flights until a sufficient volume of semen has been received from several males. Another possible organ by which queens monitor the volume of semen received is the medial and lateral oviducts because semen is deposited in these organs upon mating and is then transferred to the spermatheca. No data are available to conclude which organ is more likely to be involved in post-mating changes.

In this chapter, we introduce sexual behaviors and related physiological mechanisms in queens and males in social insects. Unique reproductive strategies may have evolved under a complicated social environment, including queen-worker conflicts over reproduction. However, many issues on the regulatory systems of sexual behaviors remain to be resolved. These should be investigated by future studies using new approaches.

REFERENCES

Akasaka, S., Sasaki, K., Harano, K., and Nagao, T. (2010) Dopamine enhances locomotor activity for mating in male honeybees (*Apis mellifera* L.). *J. Insect Physiol.*, 56, 1160-1166.

Arnold, G., Budharugsa, S. and Masson, C. (1988) Organization of the antennal lobe in the queen honey bee, *Apis mellifera* L. *J. Insect Morphol. and Embryol.*17, 185-195

Aron, S., Passera, L. and Keller, L. (1994) Queen-worker conflict over sex ratio: A comparison of primary and secondary sex ratios in the Argentine ant, *Iridomyrmex humilis*. *J. Evol. Biol.*, 7, 403-418.

Aron, S., Vargo, E. and Passera, L. (1995) Initial and secondary sex ratios in monogyne colonies of the fire ant. *Anim. Behav.*, 49, 749-757.

Akre, R. D. (1982) Social wasps. In: *Social Insects, Vol. 4* (ed. Harmann, H. R.). pp. 1-105. Academic, New York.

Baer, B., Morgan, E. D. and Schmid-Hempel, P. (2001) A nonspecific fatty acid within the bumblebee mating plug prevents females from remating. *Proc. Natl. Acad. Sci. US,* 98,3926-3928.

Beninger, R. J. (1983) The role of dopamine in locomotor activity and learning. *Brain Res. Rev.*, 6, 173-196.

Bernasconi, G., Ratnieks, F. L. W. and Rand, E. (2000) Effect of "spraying" by fighting honey bee queens (*Apis mellifera* L.) on the temporal structure of fights. *Insect. Soc.*, 47, 21-26.

Berthold, R. and Benton, A. W. (1970) Honey bee photoresponse as influenced by age. Part 2: drones and queens. *Ann. Entomol. Soc. Am.*, 63, 1113-1115.

Bloch, G., Borst, D. W., Huang, Z., Robinson, G. E., and Hefetz, A. (1996) Effects of social conditions on juvenile hormone mediated reproductive development in Bombus terrestris workers. *Physiol. Entomol.*, 21, 257-267.

Bloch, G., Borst, D. W., Huang, Z., Robinson, G. E., Cnaani, J. and Hefetz, A. (2000a) Juvenile hormone titers, juvenile hormone biosynthesis, ovarian development and social environment in *Bombus terrestris*. *J. Insect Physiol.*, 46, 47-57.

Bloch, G., Simon, T., Robinson, G. E., and Hefetz, A. (2000b) Brain biogenic amines and reproductive dominance in bumble bees (*Bombus terrestris*). *J. Comp. Physiol. A*, 186, 261-268.

Bohm, M. K. (1972) Effects of environment and juvenile hormone on ovaries of the wasp, *Polistes metricus*. *J. Insect Physiol.*, 18, 1875-1883.

Boomsma, J. J., Baer, B. and Heinze, J. (2005) The evolution of male traits in social insects. *Ann. Rev. Entomol.*, 50, 395-420.

Boulay, R., Hooper-Bui, L. M., Woodring, J. (2001) Oviposition and oogenesis in virgin fire ant females Solenopsis invicta are associated with a high level of dopamine in the brain. *Physiol. Entomol.*, 26, 294-299.

Brakefield, P. M. 1996. Seasonal polyphenism in butterflies and natural selection, *Trends Ecol. Evol.*, 11, 275–277.

Brandes, C., Sugawa, M., Menzel, R. (1990) High-performance liquid chromatography (HPLC) measurement of catecholamines in single honeybee brains reveals caste-specific differences between worker bees and queens in *Apis mellifera*. *Comp. Biochem. Physiol. C*, 97C, 53-57.

Burns, S. N., Vander Meer, R. K., and Teal, P. E. A. (2007) Mating flight activity as dealation factors for red imported fire ant (Hymenoptera: Formicidae) female alates. *Ann. Entomol. Soc. Am.*, 100, 257-264.

Carlson, N. R. (2003) *Physiology of behavior.*, 8th edn. Allyn and Bacon, Boston

Caron, D. M. and Greve, C. W. (1979) Destruction of queen cells placed in queenright *Apis mellifera* colonies. *Ann. Entomol. Soc. Am.*, 72, 405-407.

Chapman, T. (2001) Seminal fluid-mediated fitness traits in *Drosophila*. *Heredity*, 87, 511-521.

Chapuisat, M., Sundström, L. and Keller, L. (1997) Sex-ratio regulation: the economics of fratricide in ants. *Proc. R. Soc. Lond. B*, 264, 1255-1260.

Cole, B. J. (1983) Multiple mating and evolution of social behavior in the hymenoptera. *Behav. Ecol. Sociobiol.,* 12, 191-201.

Colonello, N. A. and Hartfelder, K. (2003) Protein content and pattern during mucus gland maturation and its ecdysteroid control in honey bee drones. *Apidologie*, 34, 257-267.

Colonello, N. A., Hartfelder, K. (2005) She's my girl - male accessory gland products and their function in the reproductive biology of social bees. *Apidologie*, 36, 231-244.

Cremer, S., Sledge, M. F., Heinze, J. (2002) Male ants disguised by the queen's bouquet. *Nature*, 419, 897.

Crozier, R. H. and Pamilo, P. (1996) *Evolution of social insect colonies.* Oxford University Press, Oxford, New York, Tokyo. 303 pp.

Dombroski, T. C. D., Simoes, Z. L. P. and Bitondi, M. M. G. (2003) Dietary dopamine causes ovary activation in queenless *Apis mellifera* workers. *Apidologie*, 34, 281-289.

Engels, W., Kaatz, H., Zillikens, A., Simoes, Z., Trube, A., Braun, R., and Dittrich,F. (1990) Honey bee reproduction: vitellogenin and caste-specific regulation of fertility. In: *Advances in invertebrates reproduction 5* (eds. Hoshi, M. and Yamashita, O.), pp. 495-502, Elsevier science publishers.

Engels. W., Rosenkranz, P., Adler. A., Taghizadeh, T., Luebke, G., and Francke, W. (1997) Mandibular gland volatiles and their ontogenetic patterns in queen honey bees, *Apis mellifera carnica. J. Insect Physiol.*, 43, 307-313.

Estoup, A., Solignac, M. and Cornuet, J. (1994) Precise assessment of the number of patrilines and of genetic relatedness in honeybee colonies. *Proc. R. Soc. Lond. B,* 258, 1-7.

Evans, P. D. (1980) Biogenic amines in the insect nervous system. *Adv. Insect Physiol.*, 15, 317-474.

Fahrbach, S. E., Giray, T. and Robinson, G. E. (1995) Volume changes in the mushroom bodies of adult honey bee queens. *Neurobiol. Learn. Memory*, 63, 181-191.

Fletcher, D. J. C. (1978) Vibration of queen cells by worker honeybees and its relation to the issue of swarms with virgin queens. *J. Apic. Res.*, 17, 14-26.

Fletcher, D. J. C. (1986) Triple action of queen pheromones in the regulation of reproduction in fire ant (*Solenopsis invicta*) colonies. In: *Advances in Invertebrate Reproduction 4* (eds. Porchet, M., Andries, J.-C. and Dhainaut, A.), pp. 305-316, Elsevier Science Publishers, Amsterdam.

Free, J. B. (1957) The food of adult drone honeybees (*Apis mellifera*). *Anim. Behav.*, 5, 7-11

Free, J. B., Ferguson, A. W. and Simpkins, J. R. (1992) The behaviour of queen honeybees and their attendants. *Physiol. Entomol.*, 17, 43-55.

Gary, N. E. (1962) Chemical mating attractants in the queen honey bee. *Science*, 136, 773-774.

Gary, N. E. (1963) Observations of mating behaviour in the honeybee. *J. Apic. Res.*, 2, 3-13.

Gary, N. E. and Marston, J. (1971) Mating behavior of drone honey bees with queen models. *Anim. Behav.*, 19, 299-304.

Gilley, D. C. (2001) The Behavior of Honey Bees (*Apis mellifera ligustica*) during Queen Duels. *Ethology*, 107, 601-622.

Gilley, D. C. and Tarpy, D. R. (2005) Three mechanisms of queen elimination in swarming honey bee colonies. *Apidologie*, 36, 461-474.

Gillott, C. (2003) Male accessory gland secretions: Modulators of female reproductive physiology and behavior. *Ann. Rev. Entomol.*, 48, 163-184.

Giray, T. and Robinson, G. E. (1996) Common endocrine and genetic mechanisms of behavioral development in male and worker honey bees and the evolution of division of labor. *Proc. Natl. Acad. Sci. US*, 93, 11718-11722.

Gösswald, K. and Bier, K. (1955) Beeinflussung des Geschlechtsverhältnisses durch Temperatureinwirkung bei *Formica rufa* L. *Naturwissenschaften*, 42, 133-134.

Grooters, H. J. (1987) Influences of queen piping and worker behaviour on the timing of emergence of honey bee queens. *Insect. Soc.*, 34, 181-193.

Gruntenko, N. E., Karpova, E. K., Alekseev, A. A., Chentsova, N. A., Saprykina, Z. V., Bownes, M., and Rauschenbach, I. Y. (2005) Effects of dopamine on juvenile hormone metabolism and fitness in *Drosophila virilis*. *J. Insect Physiol.*, 51, 959-968.

Harano, K. and Obara, Y. (2004a) The role of chemical and acoustical stimuli in selective queen cell destruction by virgin queens of the honeybee *Apis mellifera*. *Appl. Entmol. Zool.*, 39, 611-616.

Harano, K. and Obara, Y. (2004b) Virgin queens selectively destroy fully matured queen cells in the honeybee *Apis mellifera* L. *Insect. Soc.*, 51, 253-258.

Harano, K., Sasaki, K. and Nagao, T. (2005) Depression of brain dopamine and its metabolite after mating in European honeybee (*Apis mellifera*) queens. *Naturwissenschaften*, 92, 310-313.

Harano, K., Sasaki, M. and Sasaki, K. (2007) Effects of reproductive state on rhythmicity, locomotor activity and body weight in the European honeybee, *Apis mellifera* queens (Hymenoptera, Apini). *Sociobiol.*, 50, 189-200.

Harano, K., Shibai, Y., Sonezaki, T. and Sasaki, M. (2008a) Behavioral strategies of virgin honeybee (*Apis mellifera*) queens in sister elimination: different responses to unemerged sisters depending on maturity. *Sociobiol.*, 52, 31-46.

Harano, K., Sasaki, M., Nagao, T., and Sasaki, K. (2008b) Dopamine influences locomotor activity in honeybee queens: implications for a behavioural change after mating. *Physiol. Entomol.*, 33, 395-399.

Harano, K., Sasaki, K., Nagao, T., and Sasaki, M. (2008c) Influence of age and juvenile hormone on brain dopamine level in male honeybee (*Apis mellifera*): Association with reproductive maturation. *J. Insect Physiol.*, 54, 848-853.

Harris, J. W. and Woodring, J. (1995) Elevated brain dopamine levels associated with ovary development in queenless worker honey bees (*Apis mellifera* L.). *Comp. Biochem. Physiol.*, 111C, 271-279.

Hartfelder, K. (2000) Insect juvenile hormone: from "status quo" to high society. *Braz. J. Med. Biol. Res.*, 33, 157-177.

Haydak, M. H. (1970) Honey bee nutrition. *Ann. Rev. Entomol.*, 15, 143-156.

Heinze, J., Hölldobler, B., Yamauchi, K. (1998) Male competition in *Cardiocondyla* ants. *Behav. Ecol. Sociobiol.*, 42, 239-246.

Hölldobler, B. and Wilson, E. O. (1990) *The ants*. Belknap Press of Harvard University Press, Cambridge, Mass. 732 pp.

Hrassnigg, N., Leonhard, B. and Crailsheim, K. (2003) Free amino acids in the haemolymph of honey bee queens (*Apis mellifera* L.). *Animo Acids*, 24, 205-212.

Jaycox, E. R. (1961) The effects of various foods and temperatures on sexual maturity of the drone honey bee (*Apis mellifera*). *Ann. Entomol. Soc. Am.*, 54, 519-523.

Johnson, J. N., Hardgrave, E., Gill, C., and Moore, D. (2010) Absence of consistent diel rhythmicity in mated honey bee queen behavior. *J. Insect Physiol.*, 56, 761-773.

Julian, G. E. and Gronenberg, W. (2002) Reduction of brain volume correlates with behavioral changes in queen ants. *Brain Behav. Evol.*, 60, 152-164.

Kamakura, M. (2011) Royalactin induces queen differentiation in honeybees, *Nature*, 473, 478–483.

Keller, L., Aron, S. and Passera, L. (1996) Internest sex-ratio variation and male brood survival in the ant *Pheidole pallidula*. *Behav. Ecol.*, 7, 292-298.

Kerr, W. E. and Silveira, Z. V. (1974) A note on the formation of honeybee spermatozoa. *J. Apic. Res.*, 13, 121-126.

Kinomura, K. and Yamauchi, K. (1987) Fighting and mating behaviors of dimorphic males in the ant *Cardiocondyla wroughtoni*. *J. Ethol.*, 5, 75-81.

Kocher, S. D., Richard, F., Tarpy, D. R., and Grozinger, C. M. (2008) Genomic analysis of post-mating changes in the honey bee queen (*Apis mellifera*). *BMC Genomics* 9, 232

Kocher, S. D., Richard, F., Tarpy, D. R., and Grozinger, C. M. (2009) Queen reproductive state modulates pheromone production and queen-worker interactions in honeybees. *Behav. Ecol.*, 20, 1007-1014.

Laidlaw, H. H. and Page, R. E. (1984) Polyandry in honeybee (*Apis mellifera* L.): sperm utilization and intracolony genetic relationships. *Genetics,* 108, 985-997.

Menzel, J. G., Wunderer, H. and Stavenga, D. G. (1991) Functional morphology of the divided compound eye of the honeybee drone (*Apis mellifera*). *Tissue Cell*, 23, 525-535

Michelsen, A, Kirchner, W. H., Andersen, B. B., and Lindauer, M. (1986) The tooting and quacking vibration signals of honeybee queens: A quantitative analysis. *J. Comp. Physiol. A*, 158, 605-611.

Michener, C. D. (1974) *The Social Behavior of the Bees*. The Belknap Press of Harverd University Press, Cambridge, Mass. 404 pp.

Moors, L., Spaas, O., Koeniger, G. and Billen, J. (2005) Morphological and ultrastructural changes in the mucus glands of Apis mellifera drones during pupal development and sexual maturation. *Apidologie*, 36, 245-254.

Neckameyer, W. S. (1996) Multiple roles for dopamine in *Drosophila* development. *Dev. Biol.*, 176, 209-219.

Nomura, S., Takahashi, J., Sasaki, T., Yoshida, T. and Sasaki, M. (2009) Expression of the dopamine transporter in the brain of the honeybee, *Apis mellifera* L. (Hymenoptera: Apidae). *Appl. Entomol. Zool.*, 44, 403-411.

Ohtani, T. (1985) An ethological study of adult female honeybees within the hive. Ph.D. thesis, Hokkaido University.

Page, R. E. and Metcalf, R. A. (1982) Multiple mating, sperm utilization, and social evolution. *Am. Nat.*, 119, 263-281.

Page, R. E. and Robinson, G. E. (1991) The genetics of division of labour in honey bee colonies. *Adv. Insect. Physiol.,* 23, 117-169.

Passera, L. and Aron, S. (1996) Early sex discrimination and male brood elimination by workers of the Argentine ant. *Proc. R. Soc. Lond. B,* 263, 1041-1046.

Patricio, K. and Cruz-Landim, C. (2003) *Apis mellifera* (Hymenoptera, Apini) ovary development in queens and in workers from queenright and queenless colonies. *Sociobiol.*, 42, 771-780.

Pendleton, R. G., Robinson, N., Roychowdhury, R., Rasheed, A., and Hillman, R. (1996) Reproduction and development in *Drosophila* are dependent upon catecholamines. *Life Sci.*, 59, 2083-2091.

Pener, M. P. and Simpson, S. J. (2009) Locust phase polyphenism: an update, *Adv. Insect Physiol.*, **36**, 1–272.

Pflugfelder, J. and Koeniger, N. (2003) Fight between virgin queens (*Apis mellifera*) is initiated by contact to the dorsal abdominal surface. *Apidologie*, 34, 249-256.

Van Praagh, J. P., Ribi, W., Wehrhahn, C., and Wittmann, D. (1980) Drone bees fixate the queen with the dorsal frontal part of their compound eyes. *J. Comp. Physiol. A*, 136, 263-266.

Richard, F., Tarpy, D. R. and Grozinger, C. M. (2007) Effects of insemination quantity on honey bee queen physiology. *PLoS ONE*, 2, e980.

Robinson, G. E. (1985) Effects of a juvenile hormone analogue on honey bee foraging behavior and alarm pheromone production. *J. Insect Physiol.*, 31, 277-282.

Robinson, G., Vargo, E. (1997) Juvenile hormone in adult eusocial Hymenoptera: Gonadotropin and behavioral pacemaker. *Arch. Insect Biochem. Physiol.*, 35, 559-583.

Robinson, G. E., Strambi, C., Strambi, A., and Huang, Z. (1992) Reproduction in worker honey bees is associated with low juvenile hormone titers and rates of biosynthesis. *Gen. Comp. Endocrinol.*, 87, 471-480.

Roseler, P., Roseler, I. and Strambi, A. (1980) The activity of corpora allata in dominant and subordinated females of the wasp *Polistes gallicus*. *Insect. Soc.*, 27, 97-107.

Ruttner, F. (1956) The mating of the honeybee. *Bee World*, 37, 3-15.

Ruttner, F. (1966) The life and flight activity of drones. *Bee World*, 47, 93-100.

Sasagawa, H., Sasaki, M. and Okada, I. (1989) Hormonal control of the division of labor in adult honeybees (*Apis mellifera* L.). I. Effect of methoprene on corpora allata and hypopharyngeal gland, and its alpha-glucosidase activity. *Appl. Entomol. Zool.*, 24, 66-77.

Sasaki, K., Satoh, T. and Obara, Y. (1995) Sperm utilization by honey bee queens; DNA fingerprinting analysis. *Appl. Entomol. Zool.*, 30, 335-341.

52

Ken-ichi Harano and Ken Sasaki

Sasaki, K., Satoh, T. and Obara, Y. (1996) The honeybee queen has the potential ability to regulate the initial sex ratio. Appl. Entomol. Zool., 31, 247-254.

Sasaki, K. and Nagao, T. (2001) Distribution and levels of dopamine and its metabolites in brains of reproductive workers in honeybees. J. Insect Physiol., 47: 1205-1216.

Sasaki, K. and Obara, Y. (2001) Nutritional factors affecting the egg sex ratio adjustment by a honeybee queen. Insect. Soc., 48, 355-359.

Sasaki, K. and Harano, K., (2010) Multiple regulatory roles of dopamine in behavior and reproduction of social insects. Trends Entomol., 6, 1-13.

Sasaki, K., Kitamura, H. and Obara, Y. (2004). Discrimination of larval sex and timing of male brood elimination by workers in honeybees (Apis mellifera L.). Appl. Entomol. Zool., 39, 393-399.

Sasaki, K., Yamasaki, K. and Nagao, T. (2007) Neuro-endocrine correlates of ovarian development and egg-laying behaviors in the primitively eusocial wasp (Polistes chinensis). J. Insect Physiol., 53, 940-949.

Sasaki, K., Yamasaki, K., Tsuchida, K. and Nagao, T. (2009) Gonadotropic effects of dopamine in isolated workers of the primitively eusocial wasp, Polistes chinensis. Naturwissenschaften, 96, 625-629.

Sasaki, K., Matsuyama, S., Harano, K., and Nagao, T. (2012) Caste differences in dopamine-related substances and dopamine supply in the brains of honeybees (Apis mellifera L.). Gen. Comp. Endocrinol., 178, 46-53.

Sasaki, K., Akasaka, S., Mezawa, R., Shimada, K., and Maekawa, K. (in press) Regulation of the brain dopaminergic system by juvenile hormone in honey bee males (Apis mellifera L.). Insect Mol. Biol.

Schneider, S. S., Painter-Kurt, S. and Degrandi-Hoffman, G. (2001) The role of the vibration signal during queen competition in colonies of the honeybee, Apis mellifera. Anim. Behav., 61, 1173-1180.

Shapiro, A. M. (1976) Seasonal Polyphenism. In: Evolutionary Biology (eds. Hecht, M. K. and Steere, W. C.), Vol. 9, pp259-333, Plenum, New York.

Snodgrass, R. E. (1956) Anatomy of the honey bee. Cornel University Press, Ithaca. 334 pp.

Sundström, L. (1994) Sex ratio bias, relatedness asymmetry and queen mating frequency in ants. Nature, 367, 266-268.

Sundström, L., Chapuisat, M. and Keller, L. (1996) Conditional manipulation of sex ratios by ant workers: a test of kin selection theory. Science, 274, 993-995.

Taber, S. (1955) Sperm distribution in the spermathecae of multiple-mated queen honey bees. *J. Econ. Entomol.*, 48, 522-525.

Tanaka, E. D. and Hartfelder, K. (2004) The initial stages of oogenesis and their relation to differential fertility in the honey bee (*Apis mellifera*) castes. *Arthropod Struct. Dev.*, 33, 431-442.

Tarpy, D. and Fletcher, D. (1998) Effects of relatedness on queen competition within honey bee colonies. *Anim. Behav.*, 55, 537-543.

Tarpy, D. and Fletcher, D. (2003) 'Spraying' behavior during queen competition in honey bees. *J. Insect Behav.*, 16, 425-437.

Tarpy, D. R., Hatch, S. and Fletcher, D. J. (2000) The influence of queen age and quality during queen replacement in honeybee colonies. *Anim. Behav.*, 59, 97-101.

Tarpy, D., Nielsen, R. and Nielsen, D. (2004) A scientific note on the revised estimates of effective paternity frequency in *Apis*. *Insectes Soc.*, 51, 203-204.

Townsend, G. F. and Lucas, C. C. (1940) The chemical nature of royal jelly. *Biochem. J.*, 34, 1155-1162.

Tozetto, S. O., Rachinsky, A. and Engels, W. (1995) Reactivation of juvenile hormone synthesis in adult drones of the honey bee, *Apis mellifera carnica*. *Experientia*, 51, 945-946.

Tozetto, S. O., Rachinsky, A. and Engels, W. (1997) Juvenile hormone promotes flight activity in drones (*Apis mellifera carnica*). *Apidologie*, 28, 77-84.

Tozetto, S. O., Bitondi, M. M. G., Dallacqua, R. P., and Simões, Z. L. P. (2007) Protein profiles of testes, seminal vesicles and accessory glands of honey bee pupae and their relation to the ecdysteroid titer. *Apidologie*, 38, 1-11.

Verlinden, H., Badisco, L., Marchal, E., Wielendaele, P. V., and Broeck, J. V. (2009) Endocrinology of reproduction and phase transition in locusts, *Gen. Comp. Endocrinol.* 162, 79–92.

Wagener-Hulme ,C., Kuehn, J. C., Schulz, D. J., and Robinson, G. E. (1999) Biogenic amines and division of labor in honey bee colonies. *J. Comp. Physiol. A*, 184, 471-479.

Winston, M. L. (1987) *The biology of the honeybee*. Harvard University Press, Cambridge, Mass. 281 pp.

Woyke, J. (1962) Natural and artificial insemination of queen honeybees. *Bee World*, 43, 21-25.

Wyatt, G. R. and Davey, K. G. (1996) Cellular and molecular actions of juvenile hormone. II. Roles of juvenile hormone in adult insects. *Adv. Insect Physiol.*, 26, 1-156.

Yamane, T., Kimura, Y., Katsuhara, M., and Miyatake, T. (2008a) Female mating receptivity inhibited by injection of male-derived extracts in *Callosobruchus chinensis. J. Insect Physiol.*, 54, 501-507.

Yamane, T., Miyatake, T. and Kimura, Y. (2008b) Female mating receptivity after injection of male-derived extracts in *Callosobruchus maculatus. J. Insect Physiol.*, 54, 1522-1527.

Yamane, T. and Miyatake, T. (2010) Induction of oviposition by injection of male-derived extracts in two *Callosobruchus* species. *J. Insect Physiol.*, 56, 1783-1788.

Yoshizawa, J., Yamauchi, K. and Tsuchida, K. (2011) Decision-making conditions for intra- or inter-nest mating of winged males in the male-dimorphic ant *Cardiocondyla minutior. Insect. Soc.*, 58, 531-538.

In: Human and Animal Mating
Editors: M. Nakamura and T. Ito

ISBN: 978-1-62417-085-0
© 2013 Nova Science Publishers, Inc.

Chapter 3

PLASTICITY IN MATING PATTERNS OF A BENTHIC NEST-HOLDING FISH RELATED TO THE EFFECTS OF NEST-SITE ABUNDANCE AND SOCIAL INTERACTIONS

Takaharu Natsumeda[*]

Department of Animal-Risk Crisis Management,
Faculty of Risk and Crisis Management,
Chiba Institute of Science, Choshi, Japan

ABSTRACT

For nest-holding animals such as fishes, abundance and spatial-temporal distribution of resources essential for reproduction (e.g. nests, breeding territories) is incredibly important for understanding the demographic and evolutionary consequences of sexual selection. In this chapter, firstly I will review the plasticity of mating patterns of the Japanese fluvial sculpin (*Cottus pollux*) related to nest site abundance in their natural condition. The number of eggs males obtained in their nests was significantly different between the two study sites with different nest site abundance, which may support the plasticity of their mating patterns related to nest abundance. Larger males occupied nests earlier and nesting males were larger than non-nesting males (i.e. size-dependent reproduction) in the area with a shortage of nest sites, whereas the same

---

[*] E-mails: natsutak@hotmail.com.

trend was not apparent in the area with sufficient nest site abundance, suggesting competitive exclusion among males for nest site under low nest abundance site. Also, males inhabit under shortage nest abundance site tend to have early maturity and shorter life span than males inhabit under sufficient nest abundance site.

Secondly, I will review nest choice experiments by males and subsequent mating experiments to quantify the effects of male-male competition on nest site choice and mating success of the male sculpins under both sufficient and shortage nest-abundance conditions. Contrary to the traditional prediction regarding mating pattern plasticity in animals in relation to the changes in nest abundance (i.e. mating pattern can shift from polygynous to monogamous as nest-site abundance increased), the results indicated exclusive polygynous mating patterns for the sculpin regardless of nest abundance. In this species, size-mediated dominance and aggressive behaviour of males may disrupt nest acquisition by other conspecific males, and may consequently result in extreme variation in mating success among males even under sufficient nest-abundance conditions. Finally, I will propose several management implications for conservation of the sculpin derived from the findings of these two topics.

GENERAL INTRODUCTION

Animals could exhibit considerable variability in ecological traits under different environmental conditions such as climate, food resource abundance, and the presence of competitor and predators. For nest-holding animals, the spatial-temporal distribution of resources essential for reproduction (e.g. nests, breeding territories) can play an important role in shaping their social structure through different sexual selection pressure (Andersson 1994; Reynolds 1996; Reichard et al. 2009). In nest-holding fishes, nests are often unevenly distributed in nature, and suitable nests are often limited abundance (Breitburg 1987; Lindström 1988; Almada et al. 1994; Järvi-Laturi et al. 2008). Both sexual selection theory and life-history theory predict that such heterogeneity in nest abundance between populations should promote different mating patterns and also may result in different life-history traits relating to different sexual selection regimes (Emlen and Oring 1977; Stearns 1992; Roff 2001; Shuster and Wade 2003).

According to the tenet of sexual selection theory, growing evidence for the plasticity of sexual selection, mating patterns and breeding characteristics of nest-holding fishes related to nest site abundance have been accumulated (Mousseau and Collins 1987; Lugli et al. 1992; Forsgren et al. 1996;

Lindström and Seppä 1996; Lehtonen and Lindström 2004). For example, mating patterns of the slimy sculpins (*Cottus cognatus*) among lakes in southern Canada could be influenced by nest-rock abundance; they were exclusively polygynous where nests were limited, while they were predominantly monogamous where potential nests were much more abundant relative to male density (Mousseau and Collins 1987). Forsgren et al. (1996) showed that nest site abundance can influence the relative contribution of male-male competition and female mate choice in two populations of sand gobies (*Pomatoschistus minutus*) on Swedish west coast with a marine environment. These studies demonstrated clearly how difference in nest-site abundance between geographically separated populations resulted in different sexual selection pressure and mating patterns of nest-holding fishes.

Lehtonen and Lindström (2004), who examined the effects of nest-site abundance and nest size distribution on habitat preference in sand gobies breeding in the two different habitat (sand and rock bottom) distributed next to each other in the Northern Baltic Sea, showed that sand gobies breeding in the two habitats experience large differences in nest-site abundance and nest size distribution. Since sand gobies from the two habitats are assumed to form one opportunistically breeding population (Lehtonen and Lindström 2004), this finding suggests that the intensity of male-male competition for nests would vary even within a single population. Much recently, Takahashi (2008) showed remarkable differences in age and size at sexual maturity in males of the freshwater goby *Tridentiger brevispinis* in the same Lake (Lake Biwa). Results of these two studies suggest that different sexual selection regime related to nest site abundance would result in different mating patterns and life-history traits of nest-holding fishes even in the same population of enclosed bays and lakes.

Despite accumulating solid evidence for the plasticity of mating patterns and life-history traits among nest-holding fishes related to nest abundance in both seas and lakes, we have little knowledge of the resource-induced plasticity of life-history traits of nest-holding fishes in stream environments (but see Lugli et al. 1992). This is due in part to the difficulty of addressing rapid changes in resource abundance under spatially and temporally variable stream environments (Hildrew and Giller 1994; Allan 1994; Matthews 1998). Lugli et al. (1992), who examined the relative importance of individual and environmental factors on nest site distribution and male mating success of the freshwater goby, *Padobobius martensi* at two hill streams drain into the same river system in Italy, revealed considerable differences in nest density, male size, male mating success between the two study sizes. But unfortunately, they did not refer to the relationship between nest density and male mating success.

Recent empirical studies based on the tenets of conservation biology, on the other hand, have stressed the danger of stream fish populations being sub-divided into local groups by instream obstructions, such as culverts, weirs, road crossings, and dams (Utzinger et al. 1998; Warren and Pardew 1998; Pluym et al. 2008).

If some of these sub-divided groups within a single population are exposed to different nest site abundances, different sexual selection regimes for such resources should result in different mating patterns and life-history traits even within a single population.

The Japanese fluvial sculpin, *Cottus pollux*, a small bottom-dwelling freshwater stream fish endemic to mountain streams in Japan (large-egg type; Goto 1990), provides an opportunity to examine the relationships between nest site abundance and sexual selection regimes in their natural habitats. They are characterized as rock nesters (Balon 1975), with males maintaining the spaces beneath rocks as nests where they mate with females (Goto 1990). Similarly with other nest-holding fishes (Lindström 1988; Bisazza et al. 1989: Takahashi et al. 2001), large body size is advantageous for males of the species to acquire suitable nests through male-male competition (Natsumeda 1998a; 1998b; Natsumeda et al. 2012).

Previous studies have found that larger males breed earlier and achieve greater mating success than smaller males (Natsumeda 2005), and they often exhibit polygynous mating patterns, similarly with other congeneric species (Brown and Downhower 1982; Goto 1987; Marconato and Bisazza 1988). One preliminary study suggests that the abundance of rocks available for nests differed markedly between study sites even in the same stream (Natsumeda 1998a). So, one may wonder that the occurrence of considerable differences in the intensity of male-male competition for nest sites stems from the different nest abundance between sites.

In this chapter, I will review the two topics of the plasticity of mating patterns of the Japanese fluvial sculpin. The first is the plasticity in mating patterns of the sculpin related to nest site abundance in their natural environment. The second is the plasticity in mating patterns of the species related to social interactions among conspecific males under experimental conditions. Finally, I will propose several management implications for conservation of the sculpin derived from the findings of these two topics.

PLASTICITY IN MATING PATTERNS OF JAPANESE FLUVIAL SCULPIN RELATED TO NEST SITE ABUNDANCE

As mentioned in general introduction, this study aimed to examine the relationship between nest site abundance and mating pattern and life-history traits based on sexual selection regimes of a single population of Japanese fluvial sculpin at two sites along a stream course. Thus, it is reasonable to predict that strong competition for nest sites between males at a site with a lower abundance of nest sites should result in (1) an apparent size difference between nest-holding males and non-nesting males; (2) an exclusive polygynous mating pattern; and (3) a shorter life span for males compared to those at the site with a higher abundance of nest sites (Natsumeda et al. submitted).

Field observation was conducted at two sites (Inabe and Kochidani) along the upper reaches of the Inabe River in central Japan (35°10' N, 136°31' E). Inabe (150 m long, $7.5-10.0$ m width, 4398 m^2) is located approximately 1 km downstream of Kochidani (150 m long, $6.2-11.0$ m wide, 1176 m^2). There was a weir to control soil erosion (2 m height) at the upstream end of the Inabe site and no individuals were captured at both sites (Natsumeda 1998a), implying that the fish were sub-divided into local habitat groups by the weir. Although Kochidani had a steeper course gradient (0.4%) than Inabe (0.2%), other environmental characteristics such as stream width, water temperature, water depth and microhabitat around the nests were similar between the two study sites (Table 1). Both study sites include raceway channel-unit habitats used as spawning grounds by this species (Natsumeda 1999; 2001). Japanese fluvial sculpin is one of the common species occurring in both of the study sites (Natsumeda et al. 1997; Natsumeda 1998a).

Since almost all of the nests of the species (33 of 35, 94.3%) found in the study sites were larger than 20 cm in maximum diameter during the breeding seasons of 1990 to 1992 (T. Natsumeda, unpublished data), we examined all the rocks with a maximum diameter larger than 20 cm in the two study sites (Inabe: $n=2,154$; Kochidani: $n=3,423$) for the presence of an underlying cavity (>10 cm in depth). Natsumeda (1998a) showed that only rocks with a maximum diameter larger than 30 cm in suitable microhabitat conditions (i.e. >15 cm water depth, <40 cm/sec bottom current velocities, and granule-pebble substrata) had an underlying cavity. Thus, the rocks that satisfied these criteria were regarded as available nest rocks.

Table 1. Environmental characteristics of the two study sites in the upper reaches of Inabe River, central Japan

Environmental characteristics	Inabe	Kochidani
Stream width (m)	9.0 (7.5-10.0)	8.0 (6.2-11.0)
Gradient (%)	0.2	0.4
Elevation (m)	160	181
Water temperature (°C)	10.1 (7.3-17.2)	10.3 (8.6-14.9)
Water depth (cm)	34.8 (27.0-56.0)	33.4 (25.0-50.0)
Nest rock diameter (cm)	48.3 (24.0-154.0)	50.7 (30.0-87.0)
Water depth of nests (cm)	30.7 (15.0-44.0)	25.9 (18.0-31.0)
Bottom current velocity of nests (cm/sec)	17.3 (9.6-40.7)	19.0 (9.4-46.0)
Dominant substrate beneath the nests	pebble, granule	pebble, granule

Figures in parenthesis indicate ranges.

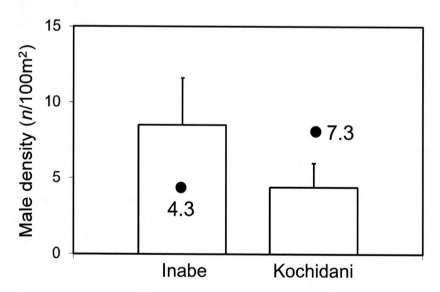

Figure 1. Densities (n/100m^2) of available rocks (closed dots with numerals) and mature males (open bars with SD) of Japanese fluvial sculpin at the two study sites (Redrawn form Natsumeda et al. submitted).

To compare the abundance of available nest rocks with that of mature males in each study site, we estimated the population number of mature males in each study site by using the equation shown by Manly and Parr (1968), Densities (n/100m^2) of available nest rocks were sufficient for mature males at Kochidani; nest rocks were in a short supply for males at Inabe (Figure 1).

Larger males of the Japanese fluvial sculpin occupied nests earlier and that nesting males were larger than non-nesting males in the area with a shortage of nest sites (Inabe). The same trend, however, was not apparent in the area with sufficient nest site abundance (Kochidani). These findings support the assumption of competitive exclusion among males for nest sites: a shortage of nest sites results in intensive male-male competition for nest sites, and males with larger body size might have an advantage in the acquisition of nests (Andersson 1994; Forsgren et al. 1996; Lindström 2001; Takahashi 2008; Natsumeda et al. 2012).

The number of eggs guarded by males at Inabe was significantly greater than the number recorded at Kochidani (Figure 2), which may partly support the plasticity of their mating patterns related to nest abundance (Mousseau and Collins 1987; Foregren et al. 1996; Lindström and Seppä 1996).

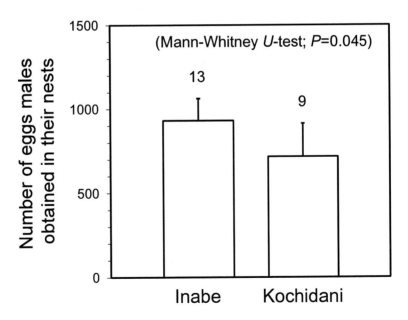

Figure 2. Comparison of the mean (with SE) number of eggs males obtained in their nests at the two study sites. Numerals in figure indicate sample size (Redrawn form Natsumeda et al. submitted).

Moreover, stepwise multiple regression analysis indicated that a model incorporating male SL, distance from the shoreline, and bottom current velocity could account for male reproductive success at Inabe (Natsumeda et al. submitted), suggesting the female preference for larger males with appropriate microhabitat choice under conditions of low nest site abundance (Natsumeda 2005). Since the inter-surface area of nests per se does not limit the number of eggs deposited by female sculpin (Natsumeda 2001), differences in the number of eggs deposited in male nests between the two sites appear to be influenced by the abundance of nest sites, as reported for slimy sculpin, *C. cognatus* (Mousseau and Collins 1987).

The age at sexual maturity of male sculpin tended to delay under sufficient nest abundance site (Natsumeda et al. submitted), suggests fewer interactions between the males for nest rocks might cause delay in the age at sexual maturity of the species. Goto (1998) also showed that the delay in age at sexual maturity of males of the river sculpin, *C. nozawae*, stemmed from the lower population density and/or fewer interactions among individuals. Similar phenomenon has reported for freshwater goby, *Tridentiger brevispinis* in Lake Biwa, Japan (Takahashi 2008).

Age specific survival rates of males matured in a given year (from age 2 to age 3 years) in area with lower nest-site abundance was considerably lower than that of area with higher nest-site abundance (Natsumeda et al. submitted). Since age-specific survival rates of females and immature males, both of which were not involved in competition for acquisition for nest sites, was not different between sites, such shorter life span of mature male sculpin under lower nest-site availability could be explained from increased stress level as a result of aggressive interactions for nest sites (Lindström 2001). Moreover, distinctly different patterns of somatic-reproductive allocation of males between habitats with different nest-site abundance may also result in considerable differences in their life span, as shown for slimy sculpin, *C. cognatus* (Mousseau et al. 1987).

It is incredibly important to identify whether such variation in mating patterns and life-history traits has a genetically basis or whether it is environmentally-induced (i.e. phenotypic plasticity; Stearns and Koella 1986; Dewitt and Scheiner 2004). Both reciprocal transfer (*C. gobio*: Mann et al. 1984) and rearing experiments (*C. hangiongensis*: Goto 1993) have provided evidence for the plasticity of phenotypic responses of such life-history traits in freshwater sculpins with varying environmental conditions. Plasticity of phenotypic responses to varying environmental conditions, however, often masks the effects of genetic response to the development of such life-history traits (Stearns and Koella 1986; Rochet et al. 2000). Further detailed studies to identify the relative influence of genetic and environmental effects on the occurrence of variation in mating patterns and life-history traits of *C. pollux* including reciprocal transfer experiments (Mann et al. 1984) and high-resolution genetic markers (e.g. microsatellite DNA, Englbrecht et al. 1999; Lamphere and Blum 2011) are requested.

SOCIALLY-CONTROLLED NEST SITE ABUNDANCE OF THE SCULPIN BY DOMINANT MALES

Sexual selection theory predicts that the intensity of male-male competition for nests should increase with the rarity of nests and consequently result in exclusive polygenous mating patterns (Emlen and Oring 1977; Shuster and Wade 2003). The potential influence of female mate choice on male mating success, by contrast, should increase with nest abundance, which may result in monogamous mating patterns (Andersson 1994; Forsgren et al. 1996). This prediction has been corroborated by several empirical studies (Mousseau and Collins 1987; Forsgren et al. 1996). Other studies also indicated that aggressive behaviour by nest owners prevented conspecific rivals from establishing nest sites near their territories, resulting in a considerable numerical gap between the sites potentially available as nests and the sites actually used as nests (Village 1983; Gauthier and Smith 1987). Thus, one may assume that male may effort to disrupt the reproduction of other nesting males to enhance their own fitness even under sufficient nest-abundance condition (Mori 1993).

To quantify the effects of male-male competition on nest site choice and mating success of the male Japanese fluvial sculpin, experiments on 5 mature males from different 5 size classes under both sufficient (5 nests) and shortage (2 nests) nest-abundance conditions were conducted (Natsumeda et al. 2012): Nest-choice experiments showed that both male size class and nest-abundance condition had significant effects on the nesting rates of males. Following the nest-choice experiments, 10 gravid females were added in the experimental tanks. Mating experiments revealed that male size, nesting rate before addition of females, and the number of courtship attempts on females were valid variables of male mating success, regardless of nest-abundance conditions (Table 2).

Table 2. Results form the final GLMM of factors affecting male mating success of male Japanese fluvial sculpin

Parameter	df	χ^2	P	$\beta \pm SE$
Intercept				-30.04 ± 13.00
Male body size	1	7.41	0.007	0.25 ± 0.12
Nesting rate	1	7.66	0.006	5.39 ± 2.83
Courtship	1	7.49	0.006	0.19 ± 0.07

(From Natsumeda et al. 2012. Journal of Ethology, 30, 239-245. With permission).

After achieving initial mating success, the largest nesting male exhibited more frequent aggressive interaction with other conspecific males than he did before obtaining eggs in his nest (Figure 3). These findings suggest that size-mediated dominance and aggressive behaviour of males may disrupt nest acquisition of other conspecific males, and may consequently result in extreme variation in mating success among males even under sufficient nest-abundance conditions. Why did the sculpin exhibit exclusive polygynous mating pattern even under sufficient nest-abundance conditions? Results of field observation indicated that larger males of the species tended to move around larger areas, including several rocks suitable as nests, before breeding (Natsumeda 2001). These findings suggest that larger males possessed larger territories than smaller ones did (e.g. the fifteen-spined stickleback *Spinachia spinachia*: Östlund 2000), and they may also control the availability of nest sites in their territories with territorial behaviour in the wild, as suggested for the Mediterranean blenny *Aidablennius sphynx* (Kraak 1996). Also, short distance between the nests used in the experimental study (0.5 m; Natsumeda et al. submitted) could have made it easier for only one male to monopolize all nests than when the distance between potential nests was longer, as for the case in the wild (14.2 m on average; Natsumeda et al. 2012). In pupfish (*Cyprinodon pecosensis*), they established a territorial breeding system in large tanks, whereas only a single dominant male controlled most of the oviposition substrate and spawned with most females (i.e. dominance hierarchy) in small tanks at low densities (Kodric-Brown 1988). These findings imply that experimental design with smaller spatial scale may also help the appearance of such exclusive dominance hierarchy more easily than experiments with larger spatial scale.

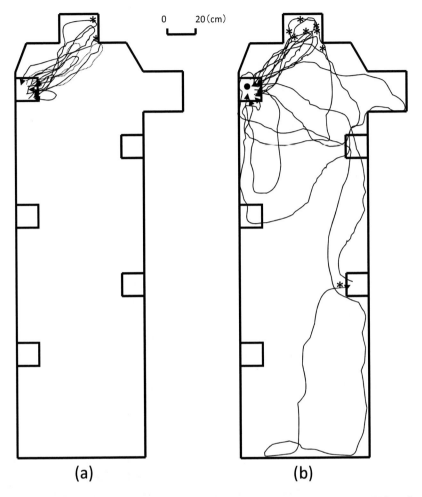

Figure 3. Behavioural traces of the largest male (fish name: T-R, 105.5 mm SL) under sufficient nest- abundance condition: (a) before spawning (20:00 — 21:00 h on 9th April 1996) and (b) just after spawning (20:00 — 21:00 h on 10th April 1996). Asterisks and dashed lines in each figure indicate the extent of aggressive behaviour against other males and courtship behaviour towards females, respectively. (From Natsumeda et al. 2012. Journal of Ethology, 30, 239-245. With permission).

Social control of reproductive activity is a widespread phenomenon for natural populations (Huntingford and Tuener 1987), and several empirical studies have indicated that such social environment can modulate the seasonal timing of maturation and reproduction of several fish species (Bushmann and Burns 1994; Rodd et al. 1997; Danylchuk and Tonn 2001; Wong et al. 2008).

In male fathead minnows (*Pimephales promelas*), for instance, the presence of larger males inhibited or delayed the reproductive activity of small, initially immature males in both the enclosure and pond experiments (Danylchuk and Tonn 2001). Despite all males of Japanese fluvial sculpin used in series experiment attained sexual maturity and they would have potential opportunity for mating with gravid females, some smaller males did not occupy nests even under sufficient nest-abundance conditions (Natsumeda et al. 2012).

The reason behind the lack of nest-retaining behaviour in smaller males even under sufficient nest-abundance condition would be partly attributable to behavioural harassment from larger males (Magnhagen and Kvarnemo 1989; Forsgren et al. 1996). In Japanese fluvial sculpin, males exhibit size-dependent bimodal reproductive patterns (i.e. large-bodied early spawners and small-bodied late spawners in one breeding season), but few males take part in both breeding activities (Natsumeda 2005). These findings imply a single catastrophic breeding attempt during their lifespan (i.e. semelparity; Stearns 1992). Smaller mature males of the species may adopt two alternative reproductive tactics: they delay reproduction until late phase of the breeding season with small body size, or postpone reproduction for another year, invest energy in growth during that time, and consequently take part in reproduction in early phase of the next breeding season with larger body size (Natsumeda 1998a; 2005; 2007). Thus, smaller males of the species had few opportunity of nesting and subsequent reproduction at which larger conspecific males were present nearby.

Such behaviourally induced limitation in nest-site abundance is also reported in several nest-holding bird species (Village 1983; Gauthier and Smith 1987; Newton 1994; Duckworth 2006). In the European kestrels *Falco finnunculus*, for instance, most of the available nests for the species remained unoccupied within 0.5 km of a nesting pair, which may result from intensive territorial defence by the pair (Village 1983). Gauthier and Smith (1987) demonstrated that pairs of the cavity nesting ducks *Bucephala albeola*, successfully expel other pairs from the water area adjacent to the nests, and may also prevent them from using other cavities near their territory. Among passerines, individuals of a larger species (e.g. starlings) can usually evict smaller ones, but can themselves be excluded by restriction of nest hole size per se (Newton 1994). These findings suggest that behaviourally induced limitation in nest-site abundance is more common among nest-holding animals.

MANAGEMENT IMPLICATIONS

The early maturity and short life span of male Japanese fluvial sculpins under condition with shortage nest site abundance (Natsumeda et al. submitted) gives insights into management implications for freshwater sculpins inhabiting stream environments. Empirical studies have indicated that the demographic characteristics of sculpins might strongly depend on local environmental conditions such as productivity and population densities (Fox 1978; Mann et al. 1984; Goto 1998). Fox (1978) demonstrated that unproductive, harsh environments resulted in early maturity, fast growth, and short life span of populations of river bullhead, *C. gobio*. Goto (1998) also noted that a habitat group of the river sculpin, *C. nozawae,* with high population density exhibited early maturity and a shorter life span than other habitat groups along the stream course. These findings imply that the early maturity of sculpins could be interpreted as a stress indicator, as shown for several fish species (Trippel 1995; Rochet et al. 2000).

An area with shortage nest site abundance was located directly below a weir used to control soil erosion. Such weirs prevent sediment flushing downstream of the weir and often result in habitat degradation for fish and aquatic insects in the lower reach of the weir (Takahashi and Higashi 1984; Mueller et al. 2011). Since bed material loading events include large-sized particles such as cobbles that could be used by sculpin for nests (Erman et al. 1988), weirs would preclude the recruitment of potential nest resources for sculpin into the lower reach of the obstruction. Moreover, at neither site did sculpin use any rocks without cavities as nests, suggesting that they did not create nest cavities by digging out the under-surface of rocks, similar to the congeneric sculpin, *C. nozawae* (Goto 1982). This suggests that availability of nests for the sculpin might be strongly dependent on the physical condition of each local habitat enclosed by weirs (Anderson 1985). To solve this problem, modification of weirs by introducing slit structures would be expected to facilitate the deposition of a sustainable supply of differently sized bed materials in the lower reach of such instream obstructions (Nakamura et al. 2006; Nakamura and Komiyama 2010).

The results of nest choice and subsequent mating experiments show that size-mediated dominance and aggressive behaviour of male Japanese fluvial sculpin may disrupt nest acquisition by other conspecific males, and may consequently result in extreme variation in mating success among males even under sufficient nest-abundance conditions. For accurate estimates for nest-site abundance of nest-holding animals, it may be important to take into account the effects of aggressive interaction on nest-site availability, in addition to traditional criteria such as nest size and environmental characteristics around the nest site (e.g. Brown and Downhower 1982; Pope and Willis 1997).

ACKNOWLEDGMENTS

The author thanks M. Yuma, Y. Nagata, S. Mori, M. Hori, T. Narita, K. Iguchi, T. Tsuruta, H. Takeshima, S. Awata, Y. Koya, D. Tahara, R. Fujii, Y. Hayakawa, H. Yanbe, and H. Matsubara for their valuable discussions; T. Ito and Y. Shimizu for valuable information during the field surveys.

REFERENCES

Allan, J. D. (1994). Stream ecology: structure and function of running waters. Dordrecht: Kluwer.

Almada V. C., Goncalves, E. J., Santos, A. J. and Baptista, C. (1994). Breeding ecology and nest aggregations in a population of *Salavia pavo* (pisces: Blenniidae) in an area where nest sites are very scarce. *Journal of Fish Biology*, 45, 819-830.

Anderson, C. S. (1985). The structure of sculpin populations along a stream size gradient. *Environmental Biology of Fishes,* 13, 93-102.

Andersson, M. (1994). Sexual selection. New Jersey: Princeton University Press.

Balon, E. K. (1975). Reproductive guilds of fishes: a proposal and definition. *Journal of the Fisheries Research Board of Canada*, 32, 821-864.

Bisazza, A., Marconato, A. and Martin, G. (1989). Male competition and female choice in *Padogobius martensi* (Pisces, Gobbidae). *Animal Behaviour,* 38, 406-413.

Breitburg, D. L. (1987). Interspecific competition and the abundance of nest sites: factors affecting sexual selection. *Ecology,* 68, 1844-1855

Brown, L. and Downhower, J. F. (1982). Polygamy in the mottled sculpins (*Cottus bairdi*) of southwestern Montana (Pisces: Cottidae). Canadian *Journal of Zoology*, 60, 1973-1980.

Bushmann, P. J. and Burns, J. R. (1994). Social control of male sexual maturation in the swordtail characin, *Corynopoma riisei. Journal of Fish Biology*, 44, 263-272.

Danylchuk, A. J. and Tonn, W. M. (2001). Effects of social structure on reproductive activity in male fathead minnows (*Pimephales promelas*). *Behavioural Ecology*, 12, 482-489.

Dewitt, T. J. and Scheiner, S. M. (2004). Phenotypic plasticity—Functional and conceptual approaches—. New York: Oxford University Press.

Duckworth, R. A. (2006). Aggressive behaviour affects selection on morphology by influencing settlement patterns in a passerine bird. *Proceedings of the Royal Society* B, 273, 1789-1795.

Emlen, S. and Oring, L. W. (1977). Ecology, sexual selection, and the evolution of mating systems. *Science,* 197, 215-223.

Englbrecht, C. C., Largiader, C. R., Haenfling, B, and Tautz, D. (1999). Isolation and characterization of polymorphic microsatellite loci in the European bullhead *Cottus gobio* L. (Osteichthyes) and their applicability to related taxa. *Molecular Ecology*, 8, 1966-1969.

Erman, D. C., Andrews, E. D, and Yoder-Williams, M. (1988). Effects of winter floods on stream fishes in the Sierra Nevada. *Canadian Journal of Fish and Aquatic Science, 45, 2195–2200.*

Forsgren, E., Kvarnemo, C. and Lindström, K. (1996). Mode of sexual selection determined by resource abundance in two sand goby populations. *Evolution,* 50, 646-654.

Fox, P. J. (1978). Preliminary observations on different re productive strategies in the bullhead (*Cottus gobio* L.) in northern and southern England. *Journal of Fish Biology*, 12, 5-11.

Gauthier, G. and Smith, J. N. M. (1987). Territorial behaviour, nest-site availability, and breeding density in buffleheads. *Journal of Animal Ecology*, 56, 171-184.

Goto, A. (1982). Reproductive behaviour of a river sculpin, *Cottus nozawae*. Japanese Journal of Ichthyology, 28, 453-457.

Goto, A. (1987). Polygyny in the river sculpin, *Cottus hangiongensis* (pisces: Cottidae), with special reference to male mating success. *Copeia,* 1987, 32-40.

Goto, A. (1990). Alternative life-history styles of Japanese freshwater sculpins revisited. *Environmental Biology of Fishes*, 28, 101-112.

Goto, A. (1993). Clinal life-history variation in the rivers sculpin *Cottus hangiongensis*: an example of phenotypic plasticity. *Japanese Journal of Ichthyology,* 39, 363-370.

Goto, A. (1998). Life history variations in the fluvial sculpin, *Cottus nozawae* (Cottidae), along the course of a small mountain stream. *Environmental Biology of Fishes*, 52, 203-212.

Hildrew, A. G. and Giller, P. S. (1994). Patchness, species interactions and disturbance in the stream benthos. In: Giller, P. S., Hildrew, A. G. and Raffaelli, D. G. (eds) Aquatic ecology: space, pattern and process. London: Blackwell, pp 21-62.

Huntingford, F. and Turner, A. (1987). Animal conflict. London: Chapman and Hall.

Järvi-Laturi, M., Lehtonen, T. K., Pampoulie, C. and Lindström, K. (2008). Parental care behaviour of sand gobies in determined by habitat related nest structure. *Behaviour*, 145, 39-50.

Kodric-Brown, A. (1988). Effects of population density, size of habitat and oviposition substrate on the breeding system of pupfish (*Cyprinodon pecosensis*). *Ethology*, 77, 28-43.

Kraak S. B. M. (1996). A quantitative description of the reproductive biology of the Mediterranean blenny *Aidablennius sphynx* (Teleostei, Blenniidae) in its natural habitat. *Environmental Biology of Fishes*, 46, 329-342.

Lamphere, B.A. and Blum, M. J. (2011). Genetic estimates of population structure and dispersal in a benthic stream fish. *Ecology of Freshwater Fish*, 21, 75-86.

Lehtonen, T. and Lindström, K. (2004). Changes in sexual selection resulting form novel habitat use in the sand goby. *Oikos*, 104, 327-335.

Lindström, K. (1988). Male-male competition for nest sites in the sand goby, *Pomatoschistus minutus*. *Oikos*, 53, 67-73.

Lindström, K. (2001). Effects of resource distribution on sexual selection and the cost of reproduction in sand gobies. *The American Naturalist*, 158, 64-74.

Lindström, K. and Seppä, T. (1996). The environmental potential for polygyny and sexual selection in the sand goby, *Pomatoschistus minutus*. *Proceedings of the Royal Society B*, 263, 1319-1323.

Lugli, M., Bobbio, L., Torricelli, P. and Gandolfi, G. (1992). Breeding ecology and male spawning success in two hill-stream populations of the freshwater goby, *Padogobius martensi*. Environmental Biology of Fishes, 35, 37-48.

Magnhagen, C. and Kvarnemo, L. (1989). Big is better: the importance of size for reproductive success in male *Pomatoschistus minutus* (Pisces, Gobiidae). *Journal of Fish Biology*, 35, 755-763.

Manly, B. F. J. and Parr, M. J. (1968). A new method of estimating population size, survivorship, and birth rate from capture-recapture data. *Proceedings and transactions of the British Entomological and Natural History Society*, 18, 81-89.

Mann, R. H. K., Mills, C. A. and Crisp, D. T. (1984). Geographical variation in the life-history tactics of some species of freshwater fishes. In: Potts, G. W. and Wootton, R. J. (eds). Fish reproduction: strategies and tactics. London: Academic Press, pp. 171-186.

Marconato, A. and Bisazza, A. (1988). Mate choice, egg cannibalism and reproductive success in the river bullhead, *Cottus gobio* L. *Journal of Fish Biology,* 33, 905-916.

Matthews, W. J. (1998). *Patterns in freshwater fish ecology.* Dordrecht: Kluwer.

Mori, S. (1993). The breeding system of the three spined stickleback *Gasterosteus aculaeatus* (forma leiura) with special reference to spatial and temporal patterns of nesting activity. *Behaviour,* 126, 97-124.

Mousseau, T. A., Collins, N. C. and Cabana, G. (1987). A comparative study of sexual selection and reproductive investment in the slimy sculpin, *Cottus cognatus. Oikos,* 51, 156-162.

Mousseau, T. A. and Collins, N. C. (1987). Polygyny and nest site abundance in the slimy sculpin (*Cottus cognatus*). *Canadian Journal of Zoology,* 65, 2827-2829.

Mueller M, Pander J, Geist J (2011) The effects of weirs on structural stream habitat and biological communities. *Journal of Applied Ecology,* 48, 1450–1461.

Nakamura, F. and Komiyama, E. (2010). A challenge to dam improvement for the protection of both salmon and human livelihood in Shiretoko, Japan's third Natural Heritage Site. *Landscape and Ecological Engineering,* 6, 143-152.

Nakamura, K., Tockner, K. and Amano, K. (2006). River and wetland restoration: lessons from Japan. *Bioscience,* 56, 419-429.

Natsumeda, T., Kimura, S. and Nagata, Y. (1997). Sexual size dimorphism, growth and maturity of the Japanese fluvial sculpin, *Cottus pollux* (large egg type) in the Inabe river, Mie prefecture, central Japan. *Ichthyological Research,* 44, 43-50.

Natsumeda, T. (1998a). Life history and reproductive ecology of the Japanese fluvial sculpin, *Cottus pollux,* with special reference to spatial-temporal availability of nest resources. Dissertation, Kyoto University.

Natsumeda, T. (1998b). Size-assortative nest choice by the Japanese fluvial sculpin in the presence of male-male competition. *Journal of Fish Biology,* 53, 33-38.

Natsumeda, T. (1999). Year-round local movements of the Japanese fluvial sculpin, *Cottus pollux* (large egg type), with special reference to the distribution of spawning nests. *Ichthyological Research*, 46, 43-48.

Natsumeda, T. (2001). Space use by the Japanese fluvial sculpin, *Cottus pollux*, related to spatio-temporal limitations in nest resources. *Environmental Biology of Fishes*, 62, 393-400.

Natsumeda, T. (2005). Biotic and abiotic influences on nest-hatching outcome in the Japanese fluvial sculpin, *Cottus pollux*. *Environmental Biology of Fishes*, 74, 349-356.

Natsumeda, T. (2007). Variation in age at first reproduction of male Japanese fluvial sculpin induced by the timing of parental reproduction. *Journal of Fish Biology*, 70, 1378-1391.

Natsumeda, T., Mori, S. and Yuma, M. (2012). Size-mediated dominance and aggressive behavior of male Japanese fluvial sculpin *Cottus pollux* (Pisces: Cottidae) reduce nest-site abundance and mating success of conspecific rivals. *Journal of Ethology*, 30, 239-245.

Newton, I. (1994). The role of nest sites in limiting the numbers of hole-nesting birds: a review. *Biological Conservation*, 70, 265-276.

Östlund, S. (2000). Are nest characters of importance when choosing a male in the fifteen-spined sticklečback (*Spinachia spinachia*)? *Behavioural Ecology and Sociobiology*, 48, 229-235.

Pluym, J. L. V., Eggleston, D. B. and Levine, J. F. (2008). Impacts of road crossings on fish movement and community structure. *Journal of Freshwater Ecology*, 23, 565-574.

Pope, K. L. and Willis, D. W. Environmental characteristics of black crappie (*Pomoxis nigromaculatus*) nesting sites in two South Dakota waters. *Ecology of Freshwater Fish*, 6, 183-189.

Reichard, M., Ondračková, M., Bryjová, A., Smith, C. and Bryja, J. (2009). Breeding resources distribution affects selection gradients on male phenotypic traits: experimental study on lifetime reproductive success in the bittering fish (*Rhodeus amarus*). *Evolution* 63, 377-390.

Reynolds, J. D. (1996). Animal breeding systems. *Trends in Ecology and Evolution*, 11, 69-73.

Rochet, M-J., Cornillon, P-A., Sabatier, R. and Pontierm D. (2000). Comparative analysis of phylogenetic and fishing effects in life history patterns of teleost fishes. *Oikos*, 91, 255-270.

Rodd, F. H., Reznick, D. N. and Sokolowski, M. B. (1997). Phenotypic plasticity in the life history traits of guppies: responses to social environment. *Ecology*, 78, 418-433.

Roff, D. A. (2001). Age and size at maturity. In: Fox, C. W., Roff, D. A. and Fairbairn, D. J. (eds) Evolutionary ecology: concepts and case studies. Oxford: Oxford University Press, pp 99-112.

Shuster, S. M. and Wade, M. J. (2003). Mating systems and strategies. New Jersey: Princeton University Press.

Stearns, S. C. (1992). The evolution of life histories. Oxford: Oxford University Press.

Stearns, S. C. and Koella, J. C. (1986). The evolution of phenotypic plasticity in life-history traits: predictions of reaction norms for age and size at maturity. *Evolution* 40, 893-913.

Takahashi, D., Kohda, M. and Yanagisawa, Y. (2001). Male–male competition for large nests as a determinant of male mating success in a Japanese stream goby, *Rhinogobius* sp. DA. *Ichthyological Research*, 48, 91-95.

Takahashi, D. (2008). Life-history variation in relation to nest site abundance in males of the freshwater goby *Tridentiger brevispinis*. *Ecology of Freshwater Fish,* 17, 71-77.

Takahashi, G. and Higashi, S. (1984). Effects of channel alternation on fish habitat. *Japanese Journal of Limnology*, 45, 178-186.

Trippel, E. A. (1995). Age at maturity as a stress indicator in fisheries. *Bioscience,* 45, 759-771.

Utzinger, J., Roth, C. and Peter, A. (1998). Effects of environmental parameters on the distribution of bullhead, *Cottus gobio*, with particular consideration of the effects of obstructions. *Journal of Applied Ecology,* 35, 882-892.

Village, A. (1983). The role of nest-site availability and territorial behaviour in limiting the breeding density of kestrels. *Journal of Animal Ecology*, 52, 337-350.

Warren, M. L. Jr. and Pardew, M. G. (1998). Road crossings as barriers to small-stream fish movement. *Transactions of the American Fisheries Society,* 127, 637-644.

Wong, M. Y. L., Munday, P. L., Buston, P. M. and Jones, G. P. (2008). Monogamy when there is potential for polygyny: tests of multiple hypotheses in a group-living fish. *Behavioural Ecology*, 19, 353-361.

In: Human and Animal Mating
Editors: M. Nakamura and T. Ito

ISBN: 978-1-62417-085-0
© 2013 Nova Science Publishers, Inc.

Chapter 4

HUMAN SEXUAL STRATEGIES: SHORT-TERM MATING AND PARENTAL CONTROL OVER MATE CHOICE

Menelaos Apostolou[*]
University of Nicosia, Nicosia, Cyprus

ABSTRACT

In the human species, individuals engage in short-term mating strategies that enable them to acquire fitness benefits from casual mates. However, because parents and children are not genetically identical, these benefits are less valuable and more costly to their parents. For this reason the latter are likely to disapprove their children engaging in casual relationships. This chapter aims to review the evidence from several studies which indicates that parents and children disagree over short-term mating strategies, with the former considering them less acceptable for the latter than the latter consider them acceptable for themselves. Moreover, parents find it more unacceptable for their daughters to engage in short-term mating than for their sons, while male children consider casual mating more acceptable than female children. Also, mothers are more disapproving than fathers of their children's short-term mating strategies, while both parents and children consider short-term mating less acceptable within marriage. The implications of these findings for interfamily conflict are also explored.

[*] E-mail: m.apostolou@gmail.com.

INTRODUCTION

When they engage in mate choices, humans follow either a long-term or a short-term mating strategy. In the former, the individual aims to attract and retain long-term partners. The evolutionary benefits of this strategy are obvious: individuals establish a situation where they can have children to whom they can make a long-term parental investment. Benefits are also derived from using short-term mating strategies that enable mate-seekers to find partners for casual mating (Buss and Schmitt, 1993).

Casual mating has several evolutionary benefits. For instance, by engaging in multiple casual relationships a man is able to increase his reproductive success without having to sacrifice considerable resources. Similarly, a woman can increase her reproductive success, by having sexual relationships with men of superior genetic quality who would not be willing to engage in a long-term relationship with her (Buss and Schmitt, 1993).

Mating in our species has a striking uniqueness: we are the only species on the planet where parents choose mates for their children (Apostolou, 2010b).

In particular, in the great majority of contemporary pre-industrial societies parents exercise considerable influence in controlling the mate choice of their daughters and sons, with arranged marriage being the most common pattern of mating (Apostolou, 2007b, 2010; Broude and Greene, 1983).

In post-industrial societies, parents exercise indirect influence over the mating decisions of their children through means such as persuasion, threats, and appeals to loyalty (Apostolou, 2011a; Perilloux, Fleischman, and Buss, 2008; Sussman, 1953).

These patterns indicate that strategic mating, which aims to attract and retain mates for their children, is part of the parental repertoire. However, parents only have a long-term strategy, that is, they aim to get spouses for their children and not casual mates. This raises the question of whether the short-term strategic choices of children comply with the long-term goals of their parents, or if the two parties are in conflict over mating strategies.

The purpose of this chapter is to discuss evidence which favors the latter case. More specifically, I will review studies which indicate that the long-term strategies of parents conflict with the short-term strategies of their children. I will start my argument by explaining the evolutionary roots of this conflict.

PARENT-OFFSPRING CONFLICT OVER MATING IN HUMANS

Parent-offspring conflict over mating strategies is the outcome of a more general parent-offspring conflict over mating where the mate choices of children conflict with the interests of their parents. Accordingly, to understand why parents and children disagree over mating strategies one needs to understand why the two disagree over mating in general.

Let me start explaining this by taking as an example two male non-relatives, A and B. Because they do not share any genetic material, the two have no genetic interests in common; so, an action that benefits one individual does not benefit the other. For instance, if A marries and has children with a woman of superior genetic quality, this will benefit him since this trait will benefit his children. That is, he is going to have healthy and attractive children who will successfully carry his genetic material to future generations. But this is not going to affect individual B as the children of A do not carry any of his genes. Consequently, the mate choices of A are of no concern to B.

Things are different however, when A and B are relatives. This is because, when the two are genetically related, the children of A do not carry only the genes of A, but also the genes of B. It follows then that the mate choices of A come to be of interest to B: if A has children with a woman of superior genetic quality, this will benefits B because his genes, which are inside A's children, will have a greater chance of surviving and being passed to future generations. It follows also that the closer the genetic relationship between A and B, the greater the interest of B in the mate choices of A because more of his genes are inside A's children.

To make this argument more specific, I am going to assume that B is the parent of A. In this case their genetic relatedness is .5, that is, B shares 50% of his genes with A. The genetic interests of the two in this scenario greatly overlap: A will have an interest in obtaining a spouse to have children with who has superior genetic quality, because this will benefit 50% of his genes as they are inside her children. This interest is shared by B as this genetic quality will benefit 25% of her genes that are inside his grandchildren. It is clear that the genetic relatedness of the two parties results in common interests and in common desires: A would like to have a spouse of superior genetic quality and B would also like A to have a spouse of superior genetic quality.

What is less clear, but equally true, is that the desires of the two parties do not completely overlap. While a genetic relatedness of .5 means that A and B have half of their genes in common, half of their genes are not in common; that is, their genetic interests are not completely identical. Consequently, a spouse of superior quality benefits 50% of the genes of A, but only 25% of the genes of B. Therefore, good genetic quality is more beneficial to A than to B. This means that both will desire good genetic quality, but A will desire it more strongly than B because this trait is more beneficial for the offspring than the parent.

When A wishes to exercise mate choice he will strive to obtain a wife who scores very highly in genetic quality and in other traits such as character, intelligence, wealth and so on. But if A does not also score very highly in each of these qualities (which is the most likely scenario), he will soon realize that compromises have to be made because prospective wives will not be willing to enter into a long-term relationship with someone of lower mate value than themselves. Accordingly, in order to get a wife with superior genetic quality A needs to give up some other traits such as wealth and social status. Consequently, A may end up with an attractive wife (with beauty being a proxy for genetic quality), but who is relatively poor and of low social status. From A's point of view this trade-off is beneficial: A loses something but at the same time gains something else.

On the other hand, if B was to exercise mate choice on behalf of A, he would also make compromises on other traits so as to get an attractive spouse for A. But the compromises are different because for B genetic quality is much less important than it is for A. Accordingly, B would compromise less on socioeconomic status but would be willing to sacrifice some genetic quality. In effect, B would most likely end up with a daughter-in-law who may not be that attractive, but who has a good socioeconomic standing. This trade-off would be beneficial from the point of view of B.

Overall, we can see that what is optimal for A is not optimal for B: the compromises that the A makes in terms of socioeconomic status are costly, but this cost is compensated for by the gains in genetic quality. These compromises are also costly for B, but the cost is not compensated for by the gains in genetic quality, as this trait is less valuable to the parent. In effect, the mate choice of A (the offspring) inflicts a cost on B (the parent) and this cost is equal to the loss of the desirable traits that B would obtain by exercising choice.

This evolutionary reasoning has been empirically tested and verified in many instances. In particular, Apostolou (2008a) employed a sample of British parents and found that good looks are preferred more in a spouse than in an in-law. Similarly, Buunk, Park and Dubbs (2008) found that individuals consider an unattractive mating candidate more unacceptable than they think their parents would. Further studies were able to replicate these findings (Apostolou, 2011b; Buunk, and Castro Solano, 2010; Park, Dubbs and Buunk, 2008, Perilloux, Fleischman, and Buss, 2011).

Genetic quality, proxied by good looks, is not the only area of parent-offspring disagreement over mating. In particular, Apostolou (2008b) found that good family background is preferred significantly more in an in-law than in a spouse. Similarly, Buunk, Park and Dubbs (2008) found that individuals consider a mating candidate who does not come from a good family background more acceptable than they think their parents would. Subsequent studies were also able to replicate this finding (Apostolou, 2011b; Buunk, and Castro Solano, 2010; Park, Dubbs and Buunk, 2008, Perilloux et al., 2011).

Moreover, research indicates that parents and offspring disagree over the religious background of a mating candidate. That is, it appears that individuals prefer similar religious background more in an in-law than in a spouse (Apostolou, 2008a; Buunk, et al., 2008; Sprecher and Chandak, 1992). Exciting personality is another quality over which parents and offspring appear to have divergent preferences: Apostolou (2008b) found that this trait is valued more in a spouse than in an in-law, and Buunk et al. (2008) found that mate seekers consider the lack of this trait and other related qualities (i.e., sense of humor, artistic ability, creativity) in a spouse much more unacceptable than their parents.

Non-overlapping in-law and mate preferences along with the tradeoffs nature of mating mandate that the mate choices of children are costly to their parents. This hypothesis was tested with the use of a budget allocation method (Apostolou, 2011c). In particular, parents were given a budget of mate points and they were asked to allocate them in a number of traits for a prospective son-in-law and a daughter-in-law. Similarly, their children were given the same budget to be allocated to the same traits, but this time in a prospective husband and in a prospective wife. It was found that children compromised on qualities such as good family background in order to get more beauty in a spouse, while their parents compromised on beauty to get more of good family background.

Overall, non-overlapping in-law and mate preferences guide children to make compromises which do not satisfy their parents. This has several implications, one being disagreement between the two parties over mating strategies.

PARENT-OFFSPRING CONFLICT OVER MATING STRATEGIES

One strategic option for a man is to find a partner, stay with her, have children with her and divert his resources to these children. This involves a considerable cost which is balanced by the fitness benefits that come from the increased probability of having children who reach sexual maturity and who can pass his genetic material to future generations (Buss, 2003; Buss and Schmitt, 1993). On the other hand, a brief sexual encounter has a small chance of producing a child that survives to sexual maturity; nevertheless, the cumulative probability of many such relationships is much higher, making short-term mating another strategic option for a man to increase his reproductive success (Buss, 2003; Buss and Schmitt, 1993).

Having many casual relationships does not increase the fitness of a woman in terms of having more children. Nevertheless, a woman can also gain fitness benefits by engaging in short-term mating. To begin with, she can exchange sex for resources that she can divert to her children. Also, a woman can establish relationships with men who can become long-term mates or who could support her in case her husband leaves her or he does not come back from hunting or war. Furthermore, men will not make long-term commitments with women of a mate value less than their own, but they will be willing to have casual sex with women of lower mate quality; thus, a woman can marry a man of similar quality to herself and seek better genes for her children in casual relationships outside marriage (Buss and Schmitt, 1993).

In addition, when parental choice is dominant, which is usually the case in most pre-industrial human societies (Apostolou, 2007b, 2010b), offspring have to subject their mate choices to the approval of their parents. As in-law and mate preferences diverge (see above), the choices of parents are not going to satisfy the preferences of their children, a likely scenario being that the latter will find themselves married to individuals who are not as beautiful as they would like and they could obtain if they themselves were exercising mate choice. Individuals can balance this loss in genetic quality by seeking good-looking individuals outside marriage. In effect, then, short-term mating strategies are also a way for both men and women to bypass parental choice (Apostolou, 2009).

This is an obvious reason why parents are likely to disagree with the short-term mating strategies of their children, but it is not the only one. In particular, adultery is a primary reason for divorce (Betzig, 1989), so an extramarital relationship can jeopardize a marriage that parents have arranged. Moreover, if a casual relationship evolves into a long-term one, this can also be damaging for parents, since a mate's traits will reflect their offspring's preferences and not their own.

Furthermore, individuals engage in short-term mating because this increases their fitness, which means that it also increases the fitness of their parents as the two are genetically related. One primary benefit of short-term mating comes from being able to access good genes (Buss and Schmitt, 1993). Such benefit, however, counts less for parents than for their children. As discussed in the previous section, this is because the coefficient of relatedness of parents to children is 0.5, but the coefficient of relatedness of grandparents to grandchildren is only 0.25. This translates into a spouse of superior genetic quality increasing the chances that 50% of an individual's genes will pass successfully to the next generation, but an in-law of superior genetic quality increasing the chances that only 25% of an individual's genes will pass successfully to the next generation. Therefore, individuals reap more genetic benefits from a casual mate of good genetic quality than their parents do.

In sum, as offspring's short-term mating is less costly and more beneficial to them than to their parents, it is predicted that the two parties will disagree over short-term mating strategies, with daughters and sons considering these as more acceptable than their parents.

Daughters vs. Sons and Mothers vs. Fathers

Females, by investing more in their offspring, become a scarce reproductive resource to which males seek access (Trivers, 1972). Consequently, by controlling their female offspring, parents can extract valuable resources from men and their families. Accordingly, there are more fitness benefits for parents controlling the mating behavior of their daughters than that of their sons (Apostolou, 2007b). In turn, this means that parents should worry more about losing control of the mating behavior of their daughters than that of their sons.

In addition, a casual sexual relationship may result in committing a daughter's parental investment (i.e., pregnancy) to a man whom her parents do not approve (Perilloux et al., 2008). Moreover, due to parental uncertainty, males place a premium on the chastity of the female (Buss, 2003), which means that the latter's short-term mating is likely to have a bigger impact on the status of her family than the former's short-term mating. Last but not least, if a short-term relationship ends in pregnancy, it is usually the father who walks away, so the burden of childrearing falls on the mother and her parents. Consequently, maternal grandparents have to shoulder a higher burden in terms of supporting their grandchild to compensate for the loss of the father; a cost that parental grandparents may not suffer (Apostolou, 2009). For these reasons, parents are expected to consider short-term mating more unacceptable when it involves a daughter than when it involves a son.

Finally, because women are more heavily involved in the rearing of their grandchildren than men, they have to shoulder a heavier burden if their offspring's casual relationships result into a grandchild without a father or mother. As the costs of short-term mating are potentially higher for grandmothers than for grandfathers, the former are likely to be more disapproving of the short-term mating strategies of their offspring.

STUDIES ON PARENT-OFFSPRING CONFLICT OVER MATING STRATEGIES

Parent-Offspring Conflict over Mating: Within-Participants Design

One study was specifically designed to test the hypotheses derived from the evolutionary framework (Apostolou, 2009). To do so it employed an instrument that aimed to measure how acceptable people consider short-term mating strategies. The procedure for constructing this instrument was as follows: A small group of participants, which included parents and younger adults, were asked to discuss the topic of short-term mating and casual relationships and the specific short-term acts which came up in the discussion were recorded. The ones most frequently reported were added into the instrument which in its final form included 12 statements and was divided in two sections. The first section consists of eight items, which aim to measure general acceptability of short-term mating (e.g., have sex without commitments), while the second section consists of four items which aim to measure acceptability of short-term mating strategies within marriage (e.g., have an extramarital affair).

This instrument was applied to a sample of British parents who were asked to rate how acceptable they consider short-term mating strategies for themselves and for their children. That is, participants had to give acceptability ratings to the acts in the instrument twice, once for themselves and once for their parents. This was possible because sexually mature individuals with children can act both as parents and as mate-seekers. In addition, this design enables a better control for alternative explanations based on social learning, and thus, it is preferable for testing evolutionary hypotheses. For instance, a design where the responses of parents were compared with the responses of their children inevitably involves a generation gap which means that any potential differences in responses between the two can be attributed to differences in socialisation in each generation.

The results from the analysis of participants' responses indicate that individuals consider short-term mating strategies more acceptable for themselves than for their children. In particular, participants rated as more acceptable for themselves to engage in short-term mating acts that they rated it for their children. This constitutes a strong indication that there is parent-offspring disagreement over mating strategies because it demonstrates that parental and mating behaviour have been optimized differently by evolutionary forces.

It was also found that these strategies were considered to be less acceptable for daughters than for sons, and more acceptable by men than by women. That is, participants gave lower rating in the acceptability of each act when they had to rate it for their daughters than when they had to rate it for their sons. In additions, fathers gave higher acceptability rating than mothers, indicating that they view short-term mating to be more acceptable. Also, short-term mating was considered more unacceptable when individuals are married than when they are single. Finally, it was found that age had a significant effect on people's preferences as older people tended to be less approving of short-term mating than younger ones, indicating that people become more conservative, at least towards casual mating, as they age.

It is important to say that the within-participants design of this study provides strong support for the evolutionary hypotheses put forward here, since the same participants differentiated their responses according to whether they rate mating strategies for themselves or for their children. Nevertheless, this design has also disadvantages one being that all participants had to be parents so the mean age in the sample inevitably increases. The results from the statistical analysis indicate that people become more conservative as they grow older, which means that short-term mating is more acceptable at a younger age. It is very likely then that the ratings for self would have been lower had the participants been younger. In effect, the results of this study probably underestimate the degree of disagreement between parents and their daughters and sons over mating strategies.

Overall, this study has provided a strong support for the hypothesis that parents and children disagree over short-term mating strategies. Its more obvious limitation is however that it did not compare the responses of parents with their actual children, a limitation which was addressed by a subsequent study which aimed to examine disagreement within families.

Parent-Offspring Conflict over Mating: Between-Participants Design

This study took place in Cyprus and involved 148 Greek-Cypriot families (Apostolou and Georgiou, 2011). In more detail, the instrument developed in the above mentioned study was also used here. In this case however, parents had to rate how acceptable they considered short-term mating acts for their children and their actual children were asked to rate how acceptable they considered the same acts for themselves. In this way, comparisons between the two sets of responses would allow us to identify whether the two parties agree or disagree over short-term mating strategies.

The results provided strong support for the hypothesis of parent-offspring disagreement over short-term mating strategies. In particular, parents considered short-term mating acts to be more unacceptable for their children than their children considered these for themselves. It is also found that disagreement between the two parties is reduced when the children are married. That is, parents and children converged in the opinion that engaging in casual mating is unacceptable when someone is married.

Furthermore, parents consider short-term mating strategies less acceptable for their daughters than for their sons. That is, parents gave lower acceptability ratings when they had to rate each act for their daughters than for their sons. In addition, mothers and fathers agree in how much they disapprove of their daughters' short-term mating; however, this was not the case with their sons, as fathers were less disapproving than mothers. Finally, participants' age was found to have a significant effect. In particular, older parents and older children gave lower acceptability rating than younger ones, indicating that age is negatively correlated with the acceptability of short-term mating strategies.

As discussed in the previous section, one limitation of this study is that it does not control for alternative explanations based on social learning or age effects. In particular, the evolutionary hypothesis put forward here is that parents have evolved to disapprove of the short-term mating strategies of their children because this enhances their fitness in terms of better control over mate choice. The difference in approval of short-term mating strategies found in this study may also be owed, however, to the age effect: as people get older they become more conservative. Therefore, the difference in ratings can be explained by the age difference (older parents versus younger children).

This limitation has been addressed by the study which was discussed first and in which employed a within-participants design and where the age variable was held constant as the same people were asked to rate the same short-term mating acts for themselves and for their children. The combination of the results of the two studies indicates that the difference in acceptability ratings between parents and children is not only due to age difference between the two but also due to evolved predispositions. The age difference most likely adds to the evolved predisposition effect to produce the observed disagreement.

PARENT-OFFSPRING CONFLICT OVER SHORT-TERM MATING STRATEGIES AND INTERFAMILY DYNAMICS

The evolutionary framework put forward here indicates that the short-term mating strategies of children impose a cost to their parents who disapprove them. Two studies that used two different research designs provided evidence in support of this hypothesis. In this section I will attempt to explore some of the implications of these findings focusing mainly on interfamily dynamics.

The most obvious implication is that parents will keep a close watch on their children and will be sensitive to any behavior that hints involvement in casual mating. Parents may also attempt to actively discourage their children from engaging in such behaviors. For instance, they may threaten them-'I will kick you out of the house if I find you with a boyfriend'- or they may advise them-'casual mating is dangerous, there are many sexual transmitted diseases around.' On the other hand, children knowing or suspecting their parents disapproval of their short-term mating strategies will try to keep it secret from them. They may also engage in manipulation, such as lying – 'I am going out with a friend (and not a boyfriend/girlfriend).' They may also attempt to present to their parents a casual mate as a long-term mate or a mate with long-term prospects.

It is unlikely however that children will be always successful in hiding their casual relationships or deceiving their parents that they are actually in a relationship with long-term prospects. Thus, fights and disputes between the two parties are likely to rise. Parents would attempt to punish their children for inappropriate mating behavior and may take precautions in preventing any future ones. For instance, they may take them away from friends that they do not like, or change them schools to prevent negative influences. Children of course are likely to react to this, perhaps attempting to run away from home or threatening to harm themselves.

Moreover, parents are more disapproving of the short-term mating strategies of their daughters than of their sons and thus they should be more disturbed if their daughters rather than their sons engage in short-term mating. Accordingly, we expect that there will be fights between parents and children over the latter's short-term mating behavior; however, the daughters-parents fights are expected to be more severe that the sons-parents ones, as parents consider their daughters' short-term mating to be a more serious breach of good conduct.

On these grounds, we can predict further that parents will guard their daughters more closely in order to prevent them from engaging in short-term mating, and they will apply punishment if they are caught doing it. For instance, among the Hupa Indians in California: 'Her [daughter] caretakers in former times spared no pains that she might remain true to the name [virgin] until her marriage. She was not allowed to be alone with a man either in the house, or out. She was told the results of wrongdoing and severely punished by beating if she were remiss' (Goddard, 1903, p. 55).

In turn, this indicates that female mate seekers will try to be secretive about their relationships so as to avoid unleashing the wrath of their parents. If asymmetrical punishment against daughters for engaging in short-term mating was a recurrent phenomenon during human evolutionary time, it might have resulted in daughters conforming more to their parents' preferences in order to reduce the cost of punishment. This could partly explain why there is less divergence in acceptability ratings between parents and daughters than between parents and sons.

In addition, as mothers are more disapproving of their short-term mating strategies of their children than fathers it can be also predicted that there will be more fights over casual mating between mothers and children than between fathers and children. Finally, since with age people become more disapproving of short-term mating strategies, it can be predicted further there should be greater conflict in cases where there is a large age difference between parents and their children than where the age gap is small. Therefore, individuals who choose to have children later in life will find themselves disagreeing more often over mating with their offspring than parents who have offspring earlier on.

To summarize, parents and children are not genetically identical which means that they have overlapping but also conflicting interests when it comes to mate choice. Parent-offspring conflict over mating leads to parent-offspring conflict over mating strategies where the short-term mating strategies of children do not meet the approval of their parents. This disagreement has several implications for family dynamics that future research needs to explore.

REFERENCES

Apostolou, M. (2007a). Elements of parental choice: the evolution of parental preferences in relation to in-law selection. *Evolutionary Psychology*, *5*, 70-83.

Apostolou, M. (2007b). Sexual selection under parental choice: The role of parents in the evolution of human mating. *Evolution and Human Behavior*, *28*, 403-409.

Apostolou, M. (2008a). Parent-offspring conflict over mating: The case of beauty. *Evolutionary Psychology*, *6*, 303-315.

Apostolou, M. (2008b). Parent-offspring conflict over mating: The case of family background. *Evolutionary Psychology*, *6*, 456-468.

Apostolou, M. (2009). Parent-offspring conflict over mating: the case of mating strategies. *Personality and Individual Differences*, *47*, 895-899.

Apostolou, M. (2010a). Parental choice: What parents want in a son-in-law and a daughter-in-law across 67 pre-industrial societies. *British Journal of Psychology*, *101*, 695-704.

Apostolou, M. (2010b). Sexual selection under parental choice in agropastoral societies. *Evolution and Human Behavior*, *31*, 39-47.

Apostolou, M. (2011a). Parental influence over mate choice in a post-industrial context. *Letters on Evolutionary Behavioral Science*, *2*, 13-15.

Apostolou, M. (2011b). Parent-offspring conflict over mating: A replication and extension study. *Journal of Integrated Social Sciences*, *2*, 13-26.

Apostolou, M. (2011b). Parent-offspring conflict over mating: Testing the tradeoffs hypothesis. *Evolutionary Psychology*, *9*, 470-495.

Apostolou, M. and Georgiou, S. (2011). Parent-offspring conflict over short-term mating strategies. *Interpersona*, *5*, 134-148.

Betzig, L. (1989). Causes of conjugal dissolution: a cross-cultural study. *Current Anthropology*, *30*, 654-676.

Broude, G. J. and Green, S. J. (1983). Cross-cultural codes on husband-wife relationships. *Ethnology*, 22, 263-280.

Buss, D. M. (2003). *The evolution of desire: Strategies of human mating* (2nd ed.). New York: Basic Books.

Buss, D. M. and Schmitt, D. P. (1993). Sexual strategies theory: An evolutionary perspective on human mating. *Psychological Review*, *100*, 204-231.

Buunk, A. P., and Castro Solano, A. (2010). Conflicting preferences of parents and offspring over criteria for a mate: A study in Argentina. *Journal of Family Psychology, 24*, 391-399.

Buunk, A. P., Park, J. H. and Dubbs, S. L. (2008). Parent-offspring conflict in mate preferences. *Review of General Psychology*, *12*, 47-62.

Goddard, P. E. (1903). Life and culture of the Hupa. *University of California Publications in American Archaeology and Ethnology*, *1*, 1-88.

Perilloux, C., Fleischman, D. S., and Buss, D. M. (2008). The daughter guarding hypothesis: Parental influence on, and emotional reaction to, offspring's mating behavior. *Evolutionary Psychology*, *6*, 217–233.

Perilloux, C., Fleischman, D. S., and Buss, D. M. (2011). Meet the parents: Parent-offspring convergence and divergence in mate preferences. *Personality and Individual Differences*, *50*, 253-258.

Sprecher, S., and Chandak, R. (1992). Attitudes about arranged marriages and dating among men and women from India. *Free Inquiry in Creative Sociology*, *20*, 59–70.

Sussman, M. B. (1953). Parental participation in mate selection and its effect upon family continuity. *Social Forces*, *1*, 76-81.

Trivers, R. L. (1972). Parental investment and sexual selection. In B. Campell (Ed.), *Sexual selection and the descent of man: 1871-1971* (pp. 136-179). Chicago: Aldine.

INDEX

S